Real Drugs in a Virtual World

Real Drugs in a Virtual World

Drug Discourse and Community Online

Edited by Edward Murguía,
Melissa Tackett-Gibson, and Ann Lessem

LEXINGTON BOOKS

A division of
ROWMAN & LITTLEFIELD PUBLISHERS, INC.
Lanham • Boulder • New York • Toronto • Plymouth, UK

LEXINGTON BOOKS

A division of Rowman & Littlefield Publishers, Inc.
A wholly owned subsidiary of The Rowman & Littlefield Publishing Group, Inc.
4501 Forbes Boulevard, Suite 200
Lanham, MD 20706

Estover Road
Plymouth PL6 7PY
United Kingdom

British Library Cataloguing in Publication Information Available

Library of Congress Cataloging-in-Publication Data

Real drugs in a virtual world : drug discourse and community online / edited by Edward
Murguía, Melissa Tackett-Gibson, and Ann Lessem.
    p. cm.
  ISBN-13: 978-0-7391-1454-4 (cloth : alk. paper)
  ISBN-10: 0-7391-1454-9 (cloth : alk. paper)
  ISBN-13: 978-0-7391-1455-1 (pbk. : alk. paper)
  ISBN-10: 0-7391-1455-7 (pbk. : alk. paper)
  1. Drug abuse—Electronic discussion groups. 2. Drug abuse—Computer network
resources. 3. Drug abuse—Prevention—Computer network resources. 4.
Communication in drug abuse prevention. 5. Drugs and mass media. 6. Internet—Social
aspects. I. Murguía, Edward. II. Tackett-Gibson, Melissa, 1968– III. Lessem, Ann,
1947–
  HV5801.R384 2007
  362.29—dc22   2006033274

Printed in the United States of America

♾™ The paper used in this publication meets the minimum requirements of American
National Standard for Information Sciences—Permanence of Paper for Printed Library
Materials, ANSI/NISO Z39.48–1992.

Table of Contents

Acknowledgements

This work was done with the assistance of the National Institute on Drug Abuse Grant Number 5 R21 DA14882-02. Our program official at NIDA whom we would like to thank was Moira O'Brien. We hope that this book repays, at least in part, the confidence that NIDA had in us to do this work. Our conclusions, though, are our own and do not necessarily reflect positions held by other researchers at NIDA.

We would like to acknowledge the Public Policy Research Institute, College of Liberal Arts, Texas AandM University, which housed our grant and under whose auspices much of the work on the grant was done. In particular, we would like to acknowledge Dr. Charles D. Johnson, Dr. James Dyer, Robert Shultze, Lenora Angel, Jim Van Beek, and Shane Spillers of PPRI for their advice and assistance during the project. Kamesha Spates, our graduate student from the Department of Sociology, also provided essential research assistance.

We also would like to acknowledge S. Patricia Garza, Office Associate, The Mexican American and U.S. Latino Research Center (MALRC) at Texas A&M University and Misael Obregon, Research Assistant, MALRC, who greatly assisted in preparing the final drafts of the manuscript for publication. Also it was a pleasure to work with Patricia Stevenson, Production Editor at Lexington Books, who contributed greatly to the elimination of manuscript errors. The book is greatly improved because of her efforts.

Introduction

Melissa Tackett-Gibson and Edward Murguia

At present, there is much concern about the impact of online drug information on youth drug use. Many assume that the Internet introduces youth and young adults to dangerous information about hazardous substances. Others, on the other hand, contend that open and free access to accurate drug use information is critical to reducing dangerous behaviors and that Internet harm reduction sites provide valuable and accurate information on drugs and drug use. This book is a collection of articles resulting from a two-year National Institute on Drug Abuse (NIDA) funded research project on Internet use, drug information and online drug-related communities. The purpose of the book is to contribute to our understanding of the role that the Internet plays in drug use and drug communication. The primary goal of the volume, by drawing attention to the topic, is to take the debate about drug use and the Internet from polemic discourse to social scientific investigation. The researchers engaged in the project used various theoretical perspectives to explore online drug use and "club drug" culture.

The Scope of the Book

Chapters range in topic from causal factors of abuse as discussed in online forums, accounts of use and risky behaviors, the assessment of the Internet's impact on face-to-face drug use discourse, the relationship between music and drug use in online communities, and the ways in which individuals assess the accuracy of online drug information. The book will not give an in-depth review of the theoretical approaches to "virtual communities." However, we do address the topic in an article on harm reduction as "community" online and in articles that focus on the ways in which online community relates to drug use issues.

Approach

Our study was exploratory and broad in range. During the course of the research project each member of the research team applied his/her expertise to the wider topic of drug use and drug communication online. Accordingly, our work highlights a variety of ways to examine drug use as a social problem and introduces several theoretical perspectives valuable to online research. Thus the articles, as a whole, provide a comprehensive introduction to the various topics associated with the study of drug communication online. The chapter that follows in this book, entitled "Club Drugs, Online Communities, and Harm Reduction Websites on the Internet," will provide the basic literature related to the general topic of "club drug" culture online. However, each article in this volume independently also will review literature and will present data related to that article's topic.

Other Studies

The following books are related thematically to this volume and would be complementary if used together either for reference or for course selection. *Cyberia: Life in the Trenches of Hyperspace* (Rushkoff 2002) is related in that is provides a thoughtful account of early rave and techno subcultures online. The book includes information about rave organizers, neo-pagans, virtual reality entrepreneurs, computer hackers, and others. It illustrates the importance of technology to this movement and lays a solid foundation for a discussion of harm reduction and drug-related communities online. *Generation Ecstasy: Into the World of Techno and Rave Culture* (Reynolds 1998) similarly examines rave culture and technology. In particular, author Simon Reynolds highlights the interrelationship of music, technology, dance, and drug use. The book explores the links between modern electronic music and rave culture's celebration of childlike escapism through dancing and euphoric drugs such as Ecstasy.

The Internet and Health Communication (Rice and Katz 2001) underscores the growing importance of the Internet as a source of health information. While much of its text examines the applications of Internet based technologies in medical practice, it also provides a good introduction to the subject of health discourse and behavior online. It presents an in-depth analysis of the changes in human communication and health care resulting from the Internet revolution. Lastly, *Doing Internet Research: Critical Issues and Methods for Examining the Net* (Jones 1999) provides a critical examination of online social research methodologies. It examines the methodological challenges of online research, such as the ethics of online discourse analysis, and the conceptualizations of space online. It would serve as a comprehensive methodological introduction to our book.

Uniquely, our book contributes to social scientific inquiry in that it is one of the first to provide empirical data related to drug use and online communication. Other researchers have examined the attributes of online communities and the use of computer mediate communication for health purposes. However, our volume is distinctive in that it uses online qualitative and quantitative methodologies to collect and to present data related to drug use levels, behaviors, and their relationship to online activity. It is our belief that this is the only major study of its type to be completed to-date. Thus, the articles in our book contribute in that they provide vital information on drug abuse vis-à-vis online communication.

References

Jones, Steven G., ed. *Doing Internet Research: Critical Issues and Methods for Examining the Net.* Thousand Oaks, CA: Sage, 1999.

Reynolds, Simon. *Generatio n Ecstasy: Into the World of Techno and Rave Culture.* Boston: Little, Brown, 1998.

Rice, Ronald E. and James E. Katz, eds. *The Internet and Health Communication: Experiences and Expectations.* Thousand Oaks, CA: Sage, 2001.

Rushkoff, Douglas. *Cyberia: Life in the Trenches of Hyperspace.* Manchester, England: Clinamen Press, 2002.

Chapter One

Club Drugs, Online Communities, and Harm Reduction Websites on the Internet

Edward Murguía, Melissa Tackett-Gibson, and Rachel Willard

Historically, there has been an association between social groups and the particular drugs groups choose to use. In this chapter, we examine the drugs of choice, "club drugs," used by those who attend raves. We next discuss the impact of new technology, the Internet, on the use of club drugs. The Internet has the potential of changing the meaning of "community," diminishing the importance of physical space and replacing some of its importance with cyberspace. Because of the great reach of the Internet, both in terms of geography and in terms of the numbers of people who visit the Internet, we believe that our study is of singular importance. On the Internet, we focus on "Harm Reduction" websites, information centers whose main function is to minimize the harm done by mood enhancing substances. Their goal is not to eliminate the use of non-legal substances, believing that use sometimes is inevitable, but rather, to minimize the damage done to individuals by mood changing substances.

Writing before the rise of the principal club drug, Ecstasy, and before raves and the Internet, Ungerleider, Thomas, and Andrysiak (1984) examined drug use during three decades of U.S. history. They found that specific drug use trends occurred within particular social and historical contexts. During the mid-1960s, heroin was used predominately in African American ghettos, prescribed opiates such as morphine and Demerol by doctors and nurses, marijuana by Mexican immigrants and by jazz musicians, and amphetamines and sleeping pills by housewives. Also in the mid-60s, psychedelics such as LSD were used for the purpose of "consciousness expansion," and marijuana came into extensive use, even by American middle class youth. Marijuana and LSD also were associated with the counter-culture and with the anti-Vietnam war movement. Twenty years later, in the mid-1980s, the drugs of particular concern had changed. By the mid-1980s, the drugs of alarm were freebasing cocaine, phencyclidine, and methaqualone (quaaludes).

Research beginning in the 1990s has worked to document drug use trends among youth involved in the rave movement. In general, the rave movement has been characterized as a youth oriented, politically liberal movement, closely associated with technology and with non-conformist/anti-commercial behavior. One version (other versions are presented later in the chapter) traces the origins of the rave scene and its techno-orientation to industrial cities where manufacturing had been in decline, such as in Detroit. Abandoned factories, once filled with the repetitive, mechanical noise of production, became sites for raves (Bopple and Knuffer 1996). Bopple and Knuffer argue that the industrial origins of the rave movement produced a subculture both primitive and technologically sophisticated. Rave music is marked by rhythmic, repetitive beats; however, the music is produced synthetically and is set to elaborate light shows, laser displays, and video montages.

Kotarba (1993), using semi-structured interviews and participant observation, developed a detailed description of the rave scene in Houston, Texas, in the early 1990s. His description parallels descriptions of raves in Great Britain and in other parts of the United States (see, for example, Adlaf and Smart (1997) who describe the Canadian rave scene.) Kotarba indicates that the customs underlying a rave comprise a "cyberpunk culture," a blend of the technical cybernetics world and underground pop culture. Rave parties tended to be large (over one thousand people) and held in warehouses. The music was techno dance music, synthesized music with a fast constant beat played at a high volume until early morning. Clothing and dance were distinctive. Clothing included oversized clothes, clunky shoes, and knit caps, and younger female ravers wore nose rings and halter-tops with water bottles on belt loops. The dancing was individualistic and not sensual, unlike dancing to hip-hop music.

Similar to the counter-culture movement of the 1960s, the rave movement is also associated with distinctive drugs and music. Researchers have found that drug use, particularly the use of Ecstasy, was expected at raves. The drug of choice for raves is MDMA or Ecstasy, a synthetic amphetamine derivative, more technically known as 3, 4 methylenedioxymethamphetamine. Ecstasy, originally developed in 1914 as an appetite suppressant, is both a stimulant and has hallucinogenic effects. The positive results of taking it include "euphoria, insightfulness, and sociability" (Adlaf and Smart 1997:193), all of which are feelings encouraged in the rave scene. Other hallucinogens and depressants such as LSD and alcohol commonly are used as well. Music and dancing enhance the high that is reached with the taking of mood enhancing substances.

Forsyth, Barnard, and McKeganey (1997), in their study of Scottish secondary schoolers, report an association, although not necessarily a causal relationship, between Ecstasy and rave music. They indicate that in the United Kingdom, as in the United States as documented by Ungerleider, et al. (1984), there generally has been an association of a particular music with a particular type of substance. In the 1920s there was an association between cocaine and jazz. Since World War II, they indicate, there have been "the beats and cannabis, mods (amphetamines and barbiturates), hippies (cannabis and LSD), punks (glue and heroin), and acid-house (LSD and MDMA)" (p.1324). They indicate that their study results demonstrate that rave music was associated with increased substance use in all substance use categories (the legal substances, tobacco, drunkenness after using alcohol and solvents, as well as all of the nine illegal drugs that they study—cannabis, psilocybin, LSD, amphetamine, cocaine, Ecstasy, heroin, buprenorpine, temazepan).

They say that a "bottom up" (p.1324) relationship probably exists between adolescent drug use and rave music; that is, it would appear that the producers of techno music are not role models for drug use since those who put the music together are not "rock stars" in a traditional sense. Rather, they indicate, there is probably a common source for drugs and rave music. As Adlaf and Smart (1997:193) state, "It would appear more likely that adolescent substance users identify themselves with rave music rather than listening to rave music encourages drug use among this age group." Both listening to rave music and using drugs are considered recreational activities by these youth.

Several adverse health affects have been associated with the type of drug use prevalent at raves. Primarily, researchers have found that Ecstasy has been associated with a general loss of appetite and, when combined with excessive exertion (such as dancing), has been associated with convulsions, collapse, heart and lung problems, kidney failure, and hypothermia. Ecstasy use is also positively correlated with the use of other drugs (Hammersley 1999), and risks of use increase for polydrug users. These risks are heightened by the fact that most users are unaware of the purity or chemical composition of the club drug (Lenton, et al. 1997). Moreover, methods of use, such as stacking (taking high dosages of drug at one time) and injection, can produce extreme reactions to the drug or expose users to infectious diseases, such as HIV and hepatitis.

Studies have also shown that there are social risks associated with club drug use. Adlaf and Smart (1997) suggest that participation in the rave/Ecstasy subculture is associated with greater amounts of delinquency. In their Canadian study, they compared rave-only attendees with attendees at rave and "bush" parties (bush parties are large outdoor parties where alcohol rather than illicit drugs are consumed), with non-attendees of either rave or bush parties. They found that 29% of rave-only attendees report committing two or more delinquent acts during the twelve months before their survey compared to only 6% of respondents who were non-attendees of either rave or bush parties, and

50.5% of those who attended both rave and bush parties. The five delinquent acts used by Adlaf and Smart were: theft greater than $50, beating someone up, breaking into property, carrying a weapon, and gang fighting.

The Internet and "Club Drugs"

While many of the studies presented explored the patterns of club drug use and their effects, few have examined the ways in which trends migrate from place to place. Similarly, while there is a substantial amount of both quantitative and qualitative data regarding drug use in specific areas, little work has compared use patterns at multiple sites. Given the use of technology among those in the rave movement, it is important to turn to a study of technology and its relationship to "place-based," local drug use.

The rave movement emerged in the late 1980s and early 1990s at the same time that the Internet was gaining widespread use. Internet technology has the capability of creating virtual communities and of disseminating information quickly and with great reach, and these result in the possibility of the club drug phenomenon becoming different than previous waves of drug use occurring before the development of the Internet. Information on the Internet, we believe, assists us not only in understanding the rave movement but also provides a critical resource for devising interventions for drug abuse in the rave context.

Gibson (1999), in his study of the Australian drug scene, indicates that the Internet was used to publicize events and to create and maintain a community of like-minded people. The Internet provided information about activities and personalities active in techno music, the music played at raves, and about rave culture more generally. Since the Internet is computer-based, it has tended to attract a socioeconomic but alienated elite rather than minorities and the poor. Gibson indicates that ravers were comfortable with technology. They were comfortable, for example, with the technology of creating new, synthetic music; accordingly, they were, and are, comfortable with Internet use.

There have been two uses of technology important to the rave movement: 1) the use of "techno" music and video production; and 2) the incorporation of new technologies and media in communications about the community. This technological savvy and affinity is also seen in the lab-produced drugs of choice in the subculture (MDMA, GHB, etc.). An informal and aggressively anti-corporate, "do-it-yourself-and-share" value system is connected to these technological and chemical practices/values. These values, networks, and systems overlap and merge with other countercultures and underground cultures, such as "hacker" culture. Mizrach (1997) notes, "[these] subcultural groups [of hackers, ravers, and modern primitives] . . . nite culture, technology, and communication."

Attributes of Online Communities

Harm Reduction and Online Communities

In this context, responses to drug use are also being transformed. Government-sponsored efforts, with their emphasis on supply and demand reduction, have been forced to adapt technologically, as evidenced by the Webslinger crackdown of 2000.[1] The philosophy of Harm Reduction—an alternative response to drug use—has also been influenced by the new trends in drug use. Here, we trace the development of four movements: a) Harm Reduction, b) rave culture, c) the Internet, and d) club drugs. The purpose is to point out their influences on each other as well as the sociological needs that all four fill.

Harm Reduction: Its birth and transformation
Since the origin of the term and its application to drug use in the early 1980s, "Harm Reduction" has been subject to a broad array of interpretations, encompassing at times both legal and illegal mood altering agents. The term has been used to describe interventions varying from law enforcement, education aimed at mitigating the effects of substances on users, and even, on occasion, to attempts to improve the positive effects of mood altering drugs. With time, however, "harm reduction" has come to be associated with interventions designed to lessen the use and the negative effects of drug use, other than regulatory policies or abstinence-only education.

The Harm Reduction movement gained momentum as a result of the AIDS epidemic. On one hand, the epidemic posed an immediate danger to injecting drug users because needle sharing was a primary means of transmission. On the other, the control of the epidemic brought about a paradigm shift in public health efforts, bringing attention to a pragmatic approach of "safer sex" and the reduction of needle sharing. The International Harm Reduction Association best articulates this resultant philosophy:

> When attempts to reduce the prevalence and frequency of risk behavior have been pursued to the limits of their apparent effectiveness, sensible public policy requires that attempts should be made to reduce the hazardousness of each remaining risk episode (International Harm Reduction Association 2002).

Another influence in the rise of the Harm Reduction movement was a growing concern about the disproportional "casualties" of the Drug War among minority communities. The Harm Reduction Coalition states that "families and communities . . . of color . . . are frequently devastated not only by addiction, but also by arrest and incarceration, the lack of available drug treatment, infectious disease, poor housing, unemployment, etc." (The Harm Reduction Coalition 2003). As a case in point, a 2002 study found that among injecting drug users, African Americans were five times as likely as whites to contract HIV through injecting drug use ("New study reports" 2002).[2] One branch of

Harm Reduction attempts to address racial and ethnic disparities by urging that minorities and those most affected by drug use and its consequences have a voice in the planning and implementation of programs.

The Harm Reduction Movement is premised on the idea that "licit and illicit drug use is part of our world," and by implication, that not all current or prospective drug users will be reached and "saved" by anti-drug campaigns. While it often encourages abstinence of drug use, its ultimate measure of success is not abstinence, but rather changes in behavior that reduce the potential damage caused by drug use. It sees drug use patterns as part of a spectrum in which certain behaviors pose greater risks than others. As one advocate of the movement states:

> Approaches such as [those proposed by Harm Reduction] are radical in that they require the acknowledgement and acceptance that some people are not ready to give up high-risk behavior. Making connections by helping them in other ways can reduce harm and open the door to further intervention (Westermeyer n.d.).

Harm Reduction strategies seek to "meet users where they're at" through education, program implementation, and advocacy. A harm reduction group, for example, might seek to implement a needle-exchange program, educate users about how to avoid overdosing, or advocate drug treatment programs. For the purposes of this study, "Harm Reduction" will be understood as activities designed to reduce drug use and its negative effects on users through means other than law enforcement or traditional anti-drug campaigns. We also define it to include both illicit and licit drug use, including cigarettes, alcohol, and even prescription medication and herbal supplements if they are used for the purpose of altering mood. Here we focus on Harm Reduction as it relates to the club drug scene. It is not accurate to say, though, that users of club drugs and other drugs comprise two entirely distinct groups. However, the distinction is useful because some generalized differences between the two groups help to elucidate Harm Reduction strategies targeted at the two groups.

As previously noted, Harm Reduction was originally associated with injecting drug users (IDUs) because of their vulnerability to HIV/AIDS. Harm Reduction strategies, accordingly, focused primarily on prevention of disease transmission through needle exchange programs and education, and, secondarily, on prevention of other undesirable effects, such as overdose. The movement took on political overtones inasmuch as it sought to decriminalize syringe acquisition and possession in order to prevent needle sharing. The movement called for greater outreach to those who were alienated from the health care system, and thus was largely street-based. In addition, although it is likely that social factors play a role in continuing drug use, the physically addictive properties of injecting drug use suggested the need to treat users as victims. Finally, while Harm Reductionists maintained that incremental change rather than abstinence was the measure of success, implicit in their rhetoric was the idea that reduction or cessation of use was the final goal.

In contrast, club drug use, although not mutually exclusive of injection of drugs, is more commonly associated with substances that are orally ingested, smoked, or snorted.[3] Therefore, needle exchange programs take a back seat to activities focusing on education regarding dosages and drug combinations as well as testing pills for impurities. Club drug users are generally younger and more middle class than injecting drug users, so the Internet presence of many of the Harm Reduction educational activities becomes very important. The Harm Reduction movement associated with club drugs tends to take on strong political overtones, sometimes arguing for legal distinctions between "hard" and "soft" drugs, or advocating policy changes such as legalization or decriminalization. Additionally, club drug related groups are often preoccupied with free speech issues as they relate to Internet-based education and discussion. Most club drug users see themselves as recreational or experimental users; accordingly "addiction" and "treatment" are less relevant issues than in injecting drug use interventions.

We need to understand Harm Reduction activities as they relate to club drug use, as well as to the controversy surrounding these activities. In order to discuss the claims of Harm Reductionists and their opponents, it is useful first to better understand club drugs in greater detail than in our introduction, provide additional information about raves that sometimes serve as environments for their use, and describe Internet sites that conduct Harm Reduction education targeting club drug users.

The rise of club drugs
For purposes of this study, club drugs will include MDMA (Ecstasy), Lysergical Acid Diethylamide (LSD), Rohypnol, Methamphetamine (speed), Ketamine, and Gamma-hydroxybutyrate (GHB). Of these drugs, LSD has the longest history of prohibition (it was made illegal in 1967), while several others of the drugs listed above have been placed on restricted drug schedules only recently. (See Table 1 below.) Rohypnol and GHB were both banned in the United States because of their use as "date rape" drugs.

Ecstasy and other drugs
As previously mentioned, no drug is as closely associated with club drugs and the American rave scene as MDMA, or Ecstasy. This drug provides an excellent case study of the club drug phenomenon. Originally developed by Merck as a slimming agent (Dowling, McDonough, and Bost 1987), MDMA gained popularity among psychotherapists for its ability to aid clients to open up and engage in self-reflection. One therapist referred to the drug as "penicillin for the soul."

In the mid-1970s, MDMA began to find popularity as a street drug, and the name "Ecstasy" came into common use. As late as 1984, Ecstasy could be purchased openly in bars with a credit card in the United States. Prohibition, however, was quick in coming. Great Britain categorized MDMA in their most restrictive class of drugs in 1977. In the United States, the Drug Enforcement Administration (DEA) used, for the first time, its power to put an emergency ban and made the use of MDMA illegal. The drug was later classified as a Schedule I (most strictly controlled) drug, preventing both popular and medical uses as well as making it difficult to conduct research on the drug (Saunders 1993).

The rising number of people in the United States who have reported trying Ecstasy is an indicator of the increasing popularity of club drugs. In 2001, 8.1 million people in the United States reported that they had tried Ecstasy at some time in their lives, up from 6.5 million in 2000. The number of current MDMA users in 2001 was 786,000. Furthermore, the increase from 1.3 million new users in 1999 to 1.9 million in 2000 was statistically significant, as were the age-specific increases among twelve to seventeen-year-olds and eighteen to twenty-five-year-olds (Substance Abuse and Mental Health Services Administration 2001). In 2002, 10.5% of high school seniors reported that they had tried MDMA in their lifetime, and 2.4% had used the drug within the last thirty days (NIDA 2003).

Similar to many of the other club drugs, MDMA appeals to youth partly because there is a perception that it is physically non-addictive and that it does not have negative side effects. This idea is strengthened by the fact that club drugs commonly are taken orally, or in ways other than injecting. Injecting drug use not only exposes the user to infections (including HIV and Hepatitis C), it also carries the stigma associated with "addictive" drug behavior.[4] The desired effects of MDMA and other club drugs include a feeling of being at peace with the world, a sense of closeness with people around them, and an enhancement of the senses (Taylor 1996). Users report taking drugs such as GHB, Ketamine, and MDMA to reach alternative levels of consciousness (a "religious" experience) or to enhance their experience of a party or rave.

The Drug Abuse Warning Network (DAWN), created to monitor changing drug trends by tracking mentions of various drugs during hospital emergency room visits, began to detect an increase in the number of patients reporting club drug use in the 1990s. Between 1994 and 2001, the number of people accessing hospital emergency rooms who mentioned MDMA use rose steadily from 253 to 5,542. Simultaneously, the number of GHB mentions increased from 56 to 3,340, and the number of Ketamine mentions increased from 19 to 679. There were also 16 Rohypnol mentions in the first half of 2002 (Office of National Drug Control Policy 2003).

It is important to note that the DAWN system was not designed to measure the damage caused by drug use. The mention of an illicit substance during the interview does not necessarily mean that the substance was the direct cause of the problem leading to the visit. As is later noted, many emergency room visits associated with MDMA were caused by ingesting dangerous combinations, taking adulterated pills, or failing to hydrate appropriately during raves. This fact is a crucial part of the argument in favor of Harm Reduction activities.

Table 1.1. Summary of Club Drug Information

Drug	Method of Administration	Year Developed	Year Banned	Effects Sought
MDMA (Ecstacy)	Oral	1912	1985	Entactogenesis, Empathogenesis, Sensory enhancement
LSD	Oral	1938	1967	Sensory enhancement
Rohypnol	Oral, snorted	1975	1995 (Schedule III)	Relaxing of inhibitions, Short term memory loss
Methamphetamine (Speed)	Oral, smoked, snorted, injected	1919	1971	Sensory enhancement
Ketamine	Oral, snorted, injected	1962	1999 (Schedule III)	Dissociation
GHB	Oral	1960	2000 (Schedule I)	Relaxing of inhibitions, Short term memory loss

The Internet

Even before the turn of the millennium, the preeminence of the Internet was obvious. In 1999, there were 63 million hosts and well over 8 billion web pages online (Biocca 2000). In 2003, it was estimated that 174.2 million people in the United States were online, and that these users spend an average of 7.22 hours per week on the Internet (Nielson 2003).

That the Internet is an ever-present part of the lives of youth is evidenced by a study by the Kaiser Family Foundation showing that 90% of teens and young adults (age fifteen to twenty-four) report that they have used the Internet and that 75% have Internet access at home (CyberAtlas 2001). The Internet is a primary source of information: 61% of older teens (aged eighteen to nineteen years) use the Internet as a resource for completing their schoolwork. It is a primary means of developing and maintaining relationships: 91% of older teens use the Internet for email (Pastore 2002). In fact, the Internet has outstripped most other sources of information: 84% of eighteen to twenty-four-year-olds say their household is more likely to go to the Internet than to a public library for information (Pastore 2000).

Of even greater significance to this study is the fact that 68% of people between the ages of fifteen and twenty-four have looked for health related information online, and 25% say that they get a substantial amount of health-related information from the Internet. In fact, the percentage of young adults looking for health information online exceeds the percentage playing games, downloading music, or chatting. Of those who have searched for health-related information, 39% have changed their behavior as a result. Of all online youth, 23% have searched the Internet for drug and alcohol-related information (CyberAtlas 2001).

The Rave

To amplify what we said previously about raves, various stories exist about the origins of the rave and its spread to the United States. In one, the arrival of Ecstasy in England from Oregon coincided with the popularization of all-night parties in the island of Ibiza. These later spread into the U.K. in the form of all-night warehouse techno dance parties (Saunders 1993). From there, they moved to San Francisco, Los Angeles, New York, and Chicago (Kotarba 1993). As indicated previously, another version holds that raves started in Detroit and other industrial cities, where the pounding beat of techno music and artificial light shows echoed the rhythms of factory production (Bopple and Knuffer 1996).

Regardless of its origins, the rave is an event with close historical connections to mechanization and its accompanying alienation. If the last two centuries were a period of increasing penetration of the panoptic eye, raves are the escape from that gaze. Observers of raves frequently note the lack of sensuality of the dancing, which is more about "letting go" than "being seen." Although DJs may achieve popularity, raves are a far cry from the concert, where the "stage" experience is central. Another key characteristic of the rave is its simultaneous inclusiveness and exclusiveness. On one hand they are inclusive events, as flyers are broadly distributed to interested parties. On the other, raves are tied to an exclusive, esoteric subculture. There is an anonymity that many find in raves that allows them to release inhibitions and feel close to other ravers. Ecstasy and other club drugs, with their "embrace the world" effects, are not necessary for the rave experience, but some participants find that they fit in well in this environment.

Harm Reduction Organizations and Activities

As previously noted, the activities of Harm Reduction organizations are shaped around specific drugs at which they are targeted. In the case of club drugs, harm reduction consists primarily of education and pill testing. Four organizations with Harm Reduction components are described here: Erowid, DanceSafe, Lycaeum, and Bluelight. Erowid was founded in 1996 to bring together information related to various psychotropic substances. Its primary Harm Reduction activity is online education. With its server in California, the Erowid site reports an average of 330,000 visits each day, with approximately 3.5 million unique visitors in 2002. Its mission is "to provide free, reliable and accurate information about psychoactive plants and chemicals." In June of 2002, the founders of Erowid participated in a conference sponsored by the National Institute on Drug Abuse (NIDA).

DanceSafe was founded in 1999, and it constitutes one of the most extensive and active networks working with Harm Reduction in the United States. In addition to the national organization which manages their website, DanceSafe consists of nearly thirty local chapters throughout the United States. DanceSafe local chapters frequently provide information tables and pill testing services at raves and other events. DanceSafe's website is highly trafficked, receiving 130,000 visits a day as early as 2000 (Gagliano 2000). On this site, one can find a photo gallery showing pills marketed as Ecstasy but containing other substances, information as to how to avoid and how to treat an overdose, and summaries of the latest research on harmful effects of drugs. Discussion groups are also a highly utilized part of DanceSafe's site. Lycaeum does not list a date of creation or amount of traffic. It is limited to online activities (i.e., no pill testing or presence at events). Finally, Bluelight's server is in the Netherlands; Bluelight administration also participated in the 2002 NIDA conference.

The novice visitor to these sites will be surprised to see the kinds of information presented on discussion boards. For example, law students debate whether the paraphernalia laws in Florida would consider the possession of testing kits to be reasonable grounds for search. Other contributors examine medical literature and create slide shows explaining the neurochemical effects of MDMA on the brain and the problem of dopamine uptake. Several provide information about the history of various psychogenic substances. One describes the Latin etymology of the website's name.

These are not sites for those who want easy answers to difficult questions. The abundant information—including personal accounts of drug experiences—depends on users capable of making their own determinations regarding information reliability. The Lycaeum warns visitors to its site that it is their responsibility to "Be skeptical; use your head; take responsibility for your actions." Accounts of drug-related experiences from users may not carry the scientific clout of peer-reviewed journals. However, even a critic of the sites' accuracy observed that empirical evidence for withdrawal symptoms associated with GHB appeared on Erowid a year before it was recognized by scientists and doctors. Websites engaging in Harm Reduction provide somewhat simplified summaries of information by drug, including special dangers, such as that of mixing substances or overdosing. That this education occurs via the Internet ensures anonymity and the ability to eliminate geographic barriers.

Harm reduction websites usually present detailed information on the administration of drugs and potential dangers. For example, Lycaeum cautions GHB users that the drug is physically addictive, and that users should be aware of the concentration of liquid solutions of GHB, as use of the substance frequently leads to overdose. In addition to education, some Harm Reduction groups, such as DanceSafe, engage in drug testing activities. Ecstasy has been said to be the "most frequently adulterated narcotic on the market" (Gagliano 2000). Impure pills frequently contain substances such as DXM (a legal cough suppressant) or PMA (paramethoxyamphetamine), both of which can be deadly in rave settings. DXM can prevent perspiration, thus causing overheating, while PMA can raise the blood pressure to dangerous levels. Because of these and other dangers, Harm Reduction practitioners have developed the use of pill testing kits. These kits use a mixture of sulfuric acid and formaldehyde on shavings from pills to provide an indication of the presence of MDMA and other drugs. While the tests cannot provide a measure of the amount of active ingredients, they do provide a warning of the presence of adulterates (Gagliano 2000).

Controversy
While many advocates of Harm Reduction argue that their goals are not incompatible with law enforcement efforts, current policy in the United States frequently casts Harm Reduction practitioners as the enemy in their fight against club drugs. A 2003 publication from the Drug Enforcement Administration (DEA) states that the promotion of harm reduction "may mislead young people into believing that Ecstasy can be taken safely." They go on to explain:

> The growth of so-called "harm reduction" organizations has contributed to [Ecstasy's] harmless reputation. Many ravers use party drugs and advocate

their use, wrongly believing that they can be used responsibly and their effects managed properly. This myth is perpetuated, in part, by private drug education and drug testing organizations that have appeared at raves over the past ten years, testing samples of illegal drugs . . . . The idea is that if you get "good" Ecstasy, uncorrupted by other chemicals, the drug is safe. . . . But the kits are not reliable, and they often give users a false sense of security. . . . Tablets often deemed "safe" by these organizations can, in fact, be deadly. Furthermore, many law enforcement agencies believe that the practices of harm reduction organizations actually encourage drug use, and some have seen a correlation between party drug overdoses and increases in the activities of those organizations (U.S. Department of Justice 2003).

The DEA's case against Harm Reduction is based on the belief, pervasive throughout American drug policy, that abstinence-based education is the only acceptable approach to reducing drug use. By extension, access to information related to drug use and the promotion of discussion about "safer" drug use is believed to encourage drug use. New Jersey Governor Christine Todd Whitman expressed this tenant in her statement that endorsement of needle exchange programs was paramount to telling children "Just Say Maybe" (Whitman 1998). The counterargument of Harm Reductionists is that abstinence-only drug education is not effective for many people. Furthermore, they argue that scare tactics and a failure to provide substantive information ensure that youth distrust official sources of drug-related information. As one organization states in reference to the "scare tactics" of mainstream anti-drug education:

This is a common approach that has never worked. In fact, it often has the opposite effect. Individuals and organizations who exaggerate or lie about the dangers of drugs end up losing the trust of young people, who may then disbelieve all warnings about the risks of drug use. The claim that marijuana is as addictive as heroin, for example, or that Ecstasy overdoses are as common as heroin overdoses, are simply not confirmed by young people's direct experiences. . . . [and as a result] they may think that similar claims about the risks of heroin are also lies. Truthful drug education is much more effective in reducing the use and abuse of drugs among teens than scare tactics (DanceSafe n.d.).

Harm Reduction groups have been accused of glossing over the harmful effects of drug use, but their participants argue that while they do not use scare tactics, they are candid about the immediate and long-term dangers to users. Harm Reduction websites frequently state, "There is no such thing as 'safe' drug use" (DanceSafe n.d. and Bluelight n.d.). The founder of DanceSafe, Emanuel Sferios, explained the importance of promoting discussion about "safer" drug use, saying, "It's not that there are good drugs and bad drugs. . . . All of them have inherent risks. People who aren't willing to abstain need factual, unbiased information about what the drugs do and how to avoid the risks" (Gagliano 2000). That Harm Reduction sites do not seek to hide information about the harmful effects of drugs can be seen, for example, from William E. White's article in Erowid, "This is your brain on dissociatives: The bad news is finally in" (Erowid 1998).

Although traditional anti-drug efforts and the Harm Reduction movement are not inherently incompatible, tensions often have arisen between the two. Several underlying questions haunt the issue. Does the kind of information made available on the web by Harm Reduction organizations encourage drug use? If it does to any degree, then how should a society weigh potential increase in drug use against potential reduction or cessation of use and saving of lives among users? Regarding the issue of encouraging drug use, some limited research is available. One study among college and medical students found that among study participants who surfed the Internet, 9% reported that it increased their likelihood of using recreational drugs, while 19% said that it decreased their likelihood of using recreational drugs (Wax and Reynolds 2000).

The United States has historically placed greater value on the prevention of use as opposed to interventions targeted at users. Part of the controversy stems from the fact that the existence of alternative responses to drug use, such as Harm Reduction, implies a doubt about the efficacy and ethics of the drug war and the way drug related education has been conducted. Also, some Harm Reduction practitioners advocate measures that vary from legalization to decriminalization of drug use.

Legal considerations
A series of congressional bills in 2000 threatened to shut down educational activities of Harm Reduction organizations. The Methamphetamine Anti-Proliferation Act, followed by the Ecstasy Anti-Proliferation Act, proposed to restrict information available via the Internet, email, or books on the production, use, or acquisition of illicit drugs. The last two proposed a ban on information regarding the use of illegal drugs, including dosages, safety precautions, and even testing for purity ("Denied Ecstasy" 2000). These bills mark a growing antipathy toward the availability of extensive, online information regarding drug use. Pino Arlacchi, the head of the United Nations Office for Drug Control and Crime Prevention, stated in the same year that there was "a lot of extremely dangerous information" on drug use available through the Internet.

In light of these changing attitudes, Harm Reduction organizations have developed disclaimers and internal rules of operation. Netherlands-based Bluelight, for example, advises visitors to its website that they are prohibited from using the site to promote drug use, to exchange information on "ongoing or future criminal activity," or to negotiate deals. Penalty for breaking the site rules includes being barred from discussions and having posts eliminated. As one forum moderator reiterates, "This is not a forum for discussing ways to circumvent the law. Do not post information on how to smuggle drugs, ship illegal shipments, hide drugs, etc. These posts will be edited or deleted." (Bluelight, "Legal statement" n.d.) Similarly, Lycaeum states, "Planning of illegal activity and discussion of commercial suppliers are not allowed." It goes on to say:

We do not recommend growing, synthesizing, purchasing, using, administering, or selling any of the chemicals, plants, animals, fungi, and preparations described herein, regardless of legality. . . . The Lycaeum does not advocate breaking any laws, and in fact you are specifically urged not to do so." (Lycaeum 2000).

As with their educational activities, Harm Reduction groups have faced legal issues associated with their pill testing kits. In the United States, pill-testing kits are not illegal by federal law, but they may be treated as drug paraphernalia by individual state laws, and their possession may be grounds for search. For example, a Florida law (877.111 (4)) includes as paraphernalia "testing equipment used, intended for use, or designed for use in **identifying**, or in analyzing the strength, effectiveness, or purity of, controlled substances." (BeenThere 2003). Harm Reduction advocates in the United States impose additional stipulations on their participants who test drugs at raves. They cannot touch a pill at any time.

Overview of study methods

Use of Multiple Methodologies

According to McConney, Rudd, and Ayres (2002), it has become common practice to use multiple methods and mixed data in order to capture better information, and to strengthen, the validity of findings. This practice, typically called *triangulation*, was used in subsequent chapters in this book. There is increased confidence in the validity of the findings when two or more measures of the same variable produce the same results. Conversely, if results diverge with different measurement methodologies, it may indicate that one or more of these measures was influenced by spurious factors that had not been anticipated, thus providing the evaluators an opportunity to examine existing data and/or to collect additional data in order to explain or reconcile divergencies. Also, the use of multiple methodologies will enable the researchers to expand the breadth and depth of the study by answering more than one type of research question.

The broader study of the Internet and drug use, as described in subsequent chapters in this book, was designed as an ethnography of youth participating in drug related chat sites. Therefore, the Internet was our primary means of data collection. To date we have used the Internet to collect data related to the use of and attitudes toward various club drugs and drug use interventions. We have "lurked" or followed the postings of participants on several boards, including Erowid, DanceSafe, and other boards. We have made an agreement with one major drug information site to begin moderating a board on the site dedicated to research questions and conversation. Ultimately, our research team developed and administered a web-based survey of discussion board participants, results of which are included in this volume.

Notes

1. The Drug Enforcement Administration's "Operation Webslinger" culminated in the arrest of individuals accused of trafficking in GHB and its derivatives via the Internet.

2. Similarly, the Drug Policy Alliance states that while "African Americans . . . only comprise 12.2% of the population and 13% of drug users, yet [they] comprise 38% of those arrested for drug offenses and 59% of those convicted for drug offenses" (Drug Policy Alliance, n.d.).

3. Although some users inject drugs (e.g., intramuscular injections of Ketamine or intravenous injections of methamphetamines), there seem to be negative associations with injecting drug use among club drug users. One user directed comments toward those who "cannot 'get over' the stigma of an injectable entheogen" (Kent 1996).

4. Of the club drugs treated here, Ketamine is the most likely to be injected. One user pointed out that it was necessary to snort 250 mg of Ketamine to achieve the same effect as an 80 mg injection (Kent 1996).

References

Adlaf, Edward M., and Reginald G. Smart. "Party Subculture or Dens of Doom? An Epidemiological Study of Rave Attendance and Drug Use Patterns among Adolescent Students." *Journal of Psychoactive Drugs* 29 (1997):193-98.

Beck, J.E. and Rosenbaum, M. *Pursuit of Ecstasy: The MDMA Experience.* Albany: SUNY Press,1994.

BeenThere. "Kits and Cops." Message posted to <http://www.bluelight.nu/vb/showthread.php?s=91087b65c32df9b438356437e3cf6227andthreadid=27373andr=13> (9 Apr. 2003).

Biocca, F. "New media technology and youth: Trends in the evolution of new media." *Journal of Adolescent Health,* 27S (2000), 22-29.

Bluelight. (n.d.) "Disclaimer." <http://www.bluelight.nu.> (5 May 2003).

———. "Legal statement and user agreement of the New Bluelight, v.1.2." Retried May 5, 2003, from http://www.bluelight.nu/about/agreement/.

Bopple, F. and Knuffer, R. *Generation XTC: Techno und Ekstase.* Berlin: Verlag Volk und Welt Gmbh, 1996.

CyberAtlas. "Young adults use net as health information source." December 11, 2001. <http://cyberatlas.internet.com/big_picture/demographics/. (20 May 2003).

DanceSafe.(n.d.) "Philosophy and Vision." <http://www.dancesafe.org/documents/about/philosophyandvision.php> (5 May 2003).

"Denied Ecstasy" (editorial). (2000, June 19). *The Michigan Daily.*

Dowling, G. P., McDonough, E. T., III, and Bost, R. O. (1987). "'Eve' and 'Ecstasy': A report of five deaths associated with the use of MDEA and MDMA." *JAMA,* 257, 1615-17.

Forsyth, Alasdair J.M. "Places and Patterns of Drug Use in the Scottish Dance Scene." *Addiction* 91(1996): 511-21.

Forsyth, Alasdair J. M., Marina Barnard, and Neil P. McKeganey. "Musical Preference as an Indicator of Adolescent Drug Use. *Addiction* 92 (1997):1317-25.

Gagliano, R. (2000, June 9) "Ecstasy Without Fear." *LA Weekly.* Retrieved May 7, 2003, from the Media Awareness Project database.

Gibson, Chris. "Subversive Sites: Rave Culture, Spatial Politics and the Internet in Sydney, Australia." *Area* 31 (1999): 19-33.

Hammersley, Richard, Jason Ditton, Iain Smith, and Emma Short. "Patterns of Ecstasy Use by Drug Users. *British Journal of Criminology* 39 (1999): 625-47.

The Harm Reduction Coalition. (n.d.) "Principles of Harm Reduction." <www. harmreduction.org/prince.html> (5 May 2003).

International Harm Reduction Association. "What is Harm Reduction?" September 18, 2002. <http://www.ihra.org> (5 May 2003).

Kent, J. The Ketamine Konundrum. *The Erowid Vaults.* (1996, June). <http://www.erowid.org/chemicals/ketamine/ketamine_info3.shtml> (15 May 2003).

Kotarba, Joseph A. "The Rave Scene in Houston, Texas: An Ethnographic Analysis." Research Brief, Texas Commission an Alcohol and Drug Abuse, Austin, TX, 1993.

Lenton, Simon, Annabel Boys, and Kath Norcross. "Raves, Drugs, and Experience: Drug Use by a Sample of People Who Attend Raves in Western Australia. *Addiction* 92 (1997):1327-37.

Lycaeum. "Disclaimer." 2000. <http://leda.lycaeum.org/Docments/Disclaimer.8706. shtml> (15 May 2003).

National Institute on Drug Abuse (NIDA). "High School and Youth Trends." January 31, 2003. <http://www.nida.nih.gov/Infofax/HSYouthtrends.html> (16 May 2003).

"New study reports alarming rates of HIV and HCV among African American and Latino American drug injectors." (2002, November 21). *PR Newswire.* Retrieved May 6, 2003, from LexisNexis Academic database.

Nielson Net Ratings. "Average Web Usage: Month of March 2003, U.S. 2003." <http://pm.netratings.com/nnpm/owa/ NRpublicreports.usagemonthly> (21 May 2003).

Office of National Drug Control Policy. "Club Drugs. Cited Drug Abuse Warning Network." April 14, 2003. <http://www.whitehousedrugpolicy.gov/drugfact/club/ind ex.html> (19 May 2003).

Pastore, Michael. "America's young adults turn to net." CyberAtlas database, demographics. April 12, 2000 <http://cyberatlas.internet.com/big_picture/> (20 May 2003).

———. (2002, January 25). "Internet key to communication among youth." CyberAtlas database. January 25, 2002 <http://cyberatlas.internet.com/big_picture/demo graphics/> (20 May 2003).

Saunders, N. *E for Ecstasy.* London, UK. 1993 <http://www.ecstasy.org/books/e4x/> (15 May 2003).

Substance Abuse and Mental Health Service Administration (SAMHSA). "2001 National Household Survey on Drug Abuse." U.S. Department on Health and Human Services. 2001. <http://www.samhsa.gov/oas/nhsda/2k1nhsda/vol1/ toc.htm#v1> (19 May 2003).

Taylor, J. M. "MDMA FAQ." 1996 updated. <http://leda.lycaeum.org/Documents/MDM A_FAQ.10549.shtml> (15 May 2003).

Ungerleider, J. Thomas, and Therese Andrysiak. "Changes in the Drug Scene: Drug Use Trends and Behavioral Patterns." *Journal of Drug Issues* XX (1984): 217-21.

U.S. Department of Justice, Drug Enforcement Administration. *Ecstasy and Predatory Drugs.* February 2003 <www.dea.gov/concern/ clubdrugs.html> (5 May 2003).

Wax, P and Reynolds, N. "Just a click away: student Internet surfing for recreational drug information" [abstract]. *Journal of Toxicology Clinical Toxicology.* 38, (2000):531.

Westermeyer, R.W. (n.d.) "Reducing Harm: A Very Good Idea." <www.habitsmart. com> (6 May 2003)

Whitman, C. T. "Needle exchange sends mixed message" [Letter to the editor]. *Trenton Times*. May 19, 1998. http://www.dogwoodcenter.org/references/Whitman98.html (6 May 2003)

Chapter Two

The Body or the Body Politic? Risk, Harm, Moral Panic and Drug Use Discourse Online

Sarah N. Gatson

As noted by Gibson in her discussion of risk (chapter 5, this volume), "Institutions such as the media and legal and scientific professions have a substantial role in the production of risk definition and management: risk knowledge can be 'changed, magnified, dramatized, or minimized within knowledge, and to that extent [risks are] particularly open to social definition and construction'" (citing Beck, 1992: 23). The concept of risk and its close associate "harm" is one that fluctuates across time, space, and cultures. When the media involved in this process of social definition/construction is itself produced in substantial part by those we have traditionally understood as the audience, we have a situation where the media itself is framed as a risk/harm potential.

This is hardly a new situation. Across the history of many societies, the printed word and the public press have been variously understood. Along with the novel, pulp fiction, and comic books, people have understood mass media forms of print in ways that range from that which is necessary to make true freedom and democracy, to locations of danger, producing degenerate thoughts leading to destructive actions. Harmful information can thus nearly as dangerous as harmful behavior. When it comes to competing information and competing sources of information (parents or the media?), the ante of anxiety is upped even further.

Anxieties over media involve anxieties over the polysemic nature of culture (see Rodman, 2003: 25; see also Heyman, 2003, specifying the "inevitable" polysemic cultural conflicts over the First Amendment). Edles writes of culture that it involves "highly organized but also open symbolic systems such as language and fashion" (2002: 6). She further argues that the symbolic meaning(s) of cultural systems "patterns action as surely as structures of a more visible [and tangible] kind" (Edles, 2002: 7). How we articulate understandings of sociocultural phenomena like computer-mediated communication (new), drug use (old, but sometimes involving new drugs), and computer-mediated communication about drug use (old wine, new bottle?) affects the ways in which we approach such phenomena.

Thus, this chapter addresses the history of the drug use Harm Reduction Movement and particularly examines the expansion of the movement to online venues. In doing so, it documents the characteristics of harm reduction communities both online and offline and critically examines how the movement of HRM to the online arena has the potential to open up the definition of HR from microlevel and public policy medical issues, to political and legal issues at the macro level. Social movements involve actions which target either extant behaviors performed by new populations, extant behaviors performed in new ways or new venues (often due to new technologies or new ways of organizing old technologies), or new behaviors made possible by new technologies. As often as social movements target material inequalities, they also often (sometimes at the same time) target groups and/or group boundaries that expose the anxiety over stratified status positions.

Edles argues, "one of the central ways that cultural systems are *structured* is into the 'sacred' and the 'profane,' . . . [and while] The sacred and the profane are most readily apparent in the religious realm . . . this basic symbolic dichotomy underlies all kinds of cultural systems as well" (Ibid; emphasis in the original). Herein, I will discuss two major approaches to sociocultural phenomena understood as problematic that encompass this sort of dichotomous approach. "Harm Reduction" and "Abstinence-Only" occur as a linked dichotomy in a moral panic approach to social problems. The social problems involved simultaneously encompass anxieties about the body and the body politic.

Often, the body politic is superimposed upon the body, and vice versa: what the body is doing, whose body it is, and who controls the body. Risky/safe stands in for bad/good. As Dr. David F. Musto argued regarding the change in the social approach to drugs since the nineteenth century, "I am persuaded that we are dealing with large changes in our perception of the boundary between our bodies and the outside world. There are risk-taking eras and risk-reduction eras, and I believe that we are in a risk-reduction era, and will be for a while" (quoted in Fish, 2000:25). In general, "Prohibitionist and harm reduction approaches to drug use differ in understandings of who drug users are, what drug problems are, and what should be done. Where drug war policies understand all use of drugs as inherently dangerous, and all illicit drug users as inevitably addicts and abusers, harm reduction policies suggest that most people can use drugs in controlled, moderate ways" (Spade, 2002: 111). Thus, we have to ask, what/whom is being risked, and what/whom is being kept safe?

Moral Panic as a Political Approach—Historical and Contemporary

Clearly, there are qualitative differences between the perceived harms of skating rinks, dancing, circulation of information about human sexual physiology, the reading of comic books, punk music, and drug use (see Beisel, 1997: 122-124; Clemens, 1998; Barker, 1984; LeBlanc, 2000: 56-58, 62-63; Jensen, and Gerber, 1998). But there is a salient connection between the thoughts, actions, and consequences involved in all of these behaviors—each one of them has at one point in the history of more than one society been subject to the social control tactics of the moral panic. Indeed, this sort of cultural understanding of such behaviors/practices is often recurring, and anxiety and panic about one of these arenas (mass media, physical recreation, substance use, sexuality) is often connected to the others. As Rodman argues, "[Moral panics] articulate a seemingly natural and logical connection between an allegedly dangerous phenomenon and others that are already known (or at least assumed) to be genuine threats" (Rodman, 2003: 20). Moral panics are not new, and they recur around media, technology, and the body/mind generally—what goes into it (e.g., drugs; information), and what is done to it (e.g., morbidity, mortality, insanity, physical/moral degeneration).

Although much anxiety surrounds it, Rodman notes that media, is "polysemic," making predictions about its effects difficult (2003:25). So too is technology, as are the meanings of actions aimed at and ingestions into the body/mind generally. This is the polysemics of culture. The moral panic does not recognize polysemy, but rather stridently emphasizes the one good way, the one truth, the one boundary line, which no one who wishes to remain inside the group must cross. In Edles's rendering of the sacred/profane dichotomy, that which is popular, mass, or low culture is always on the profane side.

In our case, there is a mass medium in which illicit, if not always illegal, substance use information is shared. This epitomizes a scenario in which the fear of the outsiders who no longer remain unseen and unheard is made manifest in our homes. Going online brings the predator into the living-room the way no mere television show ever has. Or so the moral panic would have us believe. As I suggest in chapters 7 and 9 in this volume, this reading of the situation of course encourages a unidirectional and unidimensional moral and cultural understanding of both actual and perceived harms.

Risk/Harm, Youth, and Moral Panic—"What About the Children?"

Beisel argues that while the language of moral panics often focuses on the young and innocent, the risk of harm to the minds and bodies of the young that the campaigns articulate are not primarily about those individual young people. Instead, the panic is about reproduction. On the one hand, the reproduction of

particular mores—the group boundaries of sanctioned behaviors and ideologies—and on the other the reproduction of particular families and the class positions of those families (Beisel, 1997: 3-24; 49-75). Succinctly put, "Moral cor-
ruption of children threatens more than the reproduction of parents' cherished
values. At stake for parents is the reproduction of cultural styles, habits, and
lifestyles. The transmission of culture is important because the reproduction of
social privilege involves the reproduction of social ties" (Beisel, 1997: 5). In this
way, anxiety about innocence, innocent victims, and the future is expressed.

Barker's discussion of the English moral panic over American comic
books—a type of media whose label went from "crime" to "American" to "horror" (35)—highlights this concern with class and cultural boundaries as well:

> Certainly at this time [c1952] there was an element of an inchoate anti-US feel-
> ing. . . . the phrase popularly used to describe GIs—"overpaid, oversexed, and
> over here"—still applied. These were the first times British people had met
> other than tourist class Americans. It was the "gum-chewing, pasty-faced,
> working-class Americans" who brought the comics over (1984: 26).

Whether the anti-comics campaigners used rhetoric that was "defence of heritage against [American] imperialism" or "humanistic protection of children,"
Barker identifies a rapid shift from any empirically testable link between the
particular target in this case (American or "American-style" crime and horror
comics) and social problems (class boundary-violation, juvenile delinquency,
and crime) to "a plainly unempirical and moral argument." Indeed, it was an
argument which echoes the "gateway" arguments used in a host of historical and
contemporary moral panics, "the idea that degradation is a slippery slope. It has
a logic of its own, and once a child has touched it, worse and greater addiction
follows" (Barker, 1984: 29).[1] Yet, Barker argues the shift made the campaign
have a conceptual base that was an "anti-political moral stance" (35). How can a
moral stance of this kind—one that in part centers on policy vis-à-vis youth—be
anti-political? Indeed, how can any moral stance—with the social order of mores
at its core—be such?

Again, across the moral panic landscape, there is a difference between drugs
and comics—yes, one instance of ingestion of a particular drug can be harmful
to the physical body of a particular person who ingests it; it may indeed even
directly lead to death. But we also clearly have many examples of persons, from
the ordinary to the highly prominent (at this writing, the President of the United
States), for whom *many* ingestions over time did not lead inevitably to death and
destruction, nor even to the breakdown of class reproduction.[2] As well, there has
been no empirical evidence presented that there is *more* harm (injury or death)
since the inception of the Internet due to predators of a sexual sort or otherwise,
or in the advent of new drugs or drug-using populations, or indeed in the combination of the two (i.e., computer-mediated communication regarding drug use).[3]

Thus, in linking Internet use and drug use, we have an example of "distinct moral panics [that] add up to an imposing object of extreme terror" (Rodman, 2003: 19). Specifically, "The symbolism of drugs has a quality like Velcro, and it attaches itself to current popular fears. . . . public outrage against drug use can become so intense as to spill over into other areas of concern" (Musto, quoted in Fish, 2000: 29-30; see also 33-35 on varieties of symbolism vis-à-vis individual and group identity boundaries). What is interesting about the merging of these two panics is that in each discursive position, sensational narratives of death among the young often lead the way.

"Harm Reduction"—Legal Versus Medical Connotations

While some argue that "modern concepts of harm reduction (HR) formed during the late 1970s" (see Webster, 2005; Stoker, 2001), according to Inciardi and Harrison the term itself was coined c1984 (1999). For those who advocate harm reduction in both its medical and legal permutations, HR seems to come from a place of cultural pluralism and liberalism, focusing on the potential and actual harm to particular individuals, rather than on a sense of harm to the boundaries of the group, or between groups within a society.[4]

There have been a variety of permutations and practices that have fallen under this label, and as Harcourt notes, "there is no single harm reduction model" (2002). HR is at once a set of practices, a set of policies, and a paradigm, philosophy, or principle. The sociological contribution to HR could be traced to the work of Alfred Lindesmith, whose works explicitly challenged prohibitionist drug policy for over thirty years (see especially Lindesmith, 1965). HR is often equated with a "public health approach" to a variety of social problems involving the body, from drug and alcohol use and treatment, drug use as it intersects with AIDS and Hepatitis (particularly focusing on needle-exchange and needle-bleaching kits), prostitution, abortion and abortion rights, pornography, and anti-sodomy laws (see Gilmore, 1996; Henningfield and Slade, 1998; Harcourt, 1999; Paltrow, 1999; Fish, 2000; Cooper, 2001; Blumenson, 2002; Bergeron and Kopp, 2002; Spade, 2002; Rasmussen and Benson, 2003; and Hunter, 2004).

Typically, as a paradigm or philosophy, HR is usually at its base a pragmatic approach. This is encapsulated in Webster's equation of the everyday risk calculation of crossing the street (2005) and in HR discourse as, "'The basic idea [of harm reduction] is that you have a fallback strategy for dealing with people who are engaged in behavior that can be risky or dangerous. So if you're smoking cigarettes, smoke less or don't smoke around kids or don't throw your ashes into dry timber. If you're drinking alcohol, don't drink and drive. You ride a bicycle—use a helmet. That's harm reduction'" (Nadelmann, quoted in Spade, 2002: 110). This quotation contains a concern with both reducing harm to the individual ("don't smoke so much"), to her/his close relations ("don't smoke around kids"), and to the society at large ("don't throw your ashes into dry timber") (see also Harcourt, 1999; Malkin, 2001; Bowman, 2002; MacCoun and Reuter, 2002; Rasmussen and Benson, 2003; Brown, 2004).

HR is a widely used and even legitimate (in terms of respectable debate on practices, if not on the implementation of policies, at least in any consistent way in the U.S.) discourse across medical and legal terrains. Alongside its morally ideal opposite of abstinence-only/prohibition, HR explicitly attempts to rearticulate "harm" once again. In Harcourt's discussion of late 1990s successes in voting neighborhoods dry in Chicago—resulting in 15% of neighborhoods becoming dry by 1999—he demonstrates just how polysemic "harm" can be (1999). Harcourt argues that HR was itself "carefully crafted" within an existing prohibitive harm principle, where "Harm, not morality, structures the debate" (1999: 112).

Harcourt's is clearly not a sociological definition of morality in the sense of boundary-making I have been using, but his assertion raises an interesting point. If we are arguing about harm, are we not arguing about which harms we will tolerate, to which populations, and to which values? The harms we care about reflect our moral boundaries and cannot be separated from them. If we accept Harcourt's assertion at face value, then we have to accept that caring about the violation of civil liberties does not involve a moral boundary, but that caring about drug use or prostitution does (and that caring about all three is mutually exclusive).[5] Perhaps, then, it is harm to our morals that is the underlying structure of the debate, and thus morality was always at the center.

Medicine and the law, as so ably demonstrated by Foucault, can both be said to focus in on the body, albeit in different ways (1995: 92-93; 135-69; 195-97; 203; 207; 269). They also both center around concepts of harm, medicine tending towards a focus on the individual and law tending towards a focus on the social, social boundaries in particular. Medicine, although often concerned with controlling and disciplining the body, because it often involves a pragmatic approach to alleviate existing harms, seems much more easily aligned with HR. However, because of—as Harcourt points out—the *public* health discourse, medical HR can (and has) sought to control behaviors as well as to pragmatically work with those behaviors, what we might call a "symptom" approach to "social disease."

In contrast, the legal arena is first and foremost about the boundaries of the group, "the injury that a crime inflicts upon the social body is the disorder that it introduces into it: the scandal that it gives rise to, the example that it gives, the incitement to repeat it if it is not punished, the possibility of becoming widespread that it bears within it. . . . The penalty must have its most intense effects on those who have not committed the crime" (Foucault, 1995: 93; 95). Although the concept of the penitent/penitentiary instead gave rise to a criminal justice system that removes the public spectacle and focuses instead upon the intense reformation of individual minds while exercising extensive controls over the body and its activities (Foucault, 1995: 121-31), the general purview of the law is the group-defined boundary.

We can see these very general opposite tendencies played out in contemporary U.S. policy arenas, indeed two competing arenas of the state. On the one hand, governmental agencies/organizations like the NIH, NIDA, and the CDC all tend to take a pragmatic and explicitly public health approach, one that often also uses HR discourse. NIDA itself has interacted directly with prominent members of the HR activist community, a group that ostensibly is opposed to its mission (Fire Erowid, 2002: 2; Gatson, chapter 9, this volume). Further, as of October 2004, NIDA officially supports needle exchange policy (http://www.ihra.net/), an approach explicitly dismissed by the current president's administration vis-à-vis the AIDS epidemic.

In terms of a variety of health and public health issues, the NIH has recently come under fire specifically for much of its research into human sexuality. For example, the work of a team of researchers focuses on,

> How [do] . . . networks of people—in particular, long-haul truck drivers—work together to accelerate the spread of an epidemic, even when some of them don't know one another? Epidemiologists have long connected the spread of HIV in sub-Saharan Africa with truckers who get infected on the road and bring the virus home to their wives and girlfriends. But does the same hold true in the United States? (Yeoman, 2004)

In protesting this research, the Traditional Values Coalition reflected a discursive position that locates the populations and behaviors under study as things that should remain hidden and unknowable, "'What plausible defense can be constructed for "investigating" the sexual practices of prostitutes who service truckers?'" (Ibid).[6]

Although NIDA has long used the "drug abuse" discourse rather than embracing wholesale the drug use continuum of the most radical HR discourse, if we contrast NIDA with the DEA, we increasingly find two arms of the state that are diametrically opposed in terms of mission and policy recommendations.[7] As noted, from the outset, NIDA is not about punishment, but about medical intervention. The DEA has instead an enforcement mandate from the Department of Justice, and disseminates information upon, and enacts, moral boundaries first and foremost (See Boyd and Hutchinson, 2003, for an extended debate between the ACLU and the DEA).

However, as the medical arena is also concerned with the health of the public, the society as a whole, so the legal arena is also concerned with the individual and that concern extends beyond controlling her/him. While social versus individual connotations (public health versus informed consent in risky use) (see Gibson, Chapter 5, Murguía and Gibson, chapter 1, this volume) of harm incorporate disease and treatment models on a regular basis in medical discourse, in the legal arena there is much more of a disjuncture between HR and punishment/control. However, it remains the case in the U.S. that our notions of liberty are embedded in our legal culture, and liberty focuses on the individual.

One excellent example of an HR approach in the legal arena involves the advent of Drug Treatment Courts. The first was established in Dade County, Florida in 1989, and now there are about 425 (Goldkamp, 2000: 923). Goldkamp explores the proliferation of "problem-of-the-moment" specialty courts (e.g., family court, juvenile court), and gives an overview of the principles underlying the drug court in particular,

> the drug court model potentially represents the first stages of a fundamental paradigm shift in justice away from a predominantly punitive orientation (a.k.a. "justice as usual") toward an approach that seeks to confront and meliorate the problems associated with persons who appear in the criminal caseload. The challenges implicit in this approach are fundamental and draw into the criminal court setting expertise from health and behavioral sciences as well as linkages with a variety of social services in relationships and configurations that produce a new mix of values, aspirations and methods to guide the judicial process. To proponents, drug courts represent a major and promising departure from what had become an unrewarding routine of processing, punishing and re-punishing drug offenders to little avail. Instead, the drug court model takes on "root causes" of crime more easily ignored or viewed as someone else's responsibility. To these proponents, the drug court model signals an important shift in philosophy from punishment for drug-involved offenders to helping offenders involved with drugs. (Goldkamp, 2000: 924-925; see also Bowman, 2002)

A shift from punishment to helping does seem to point toward the predominant HR paradigm.

The legal HR discourse also focuses on other ways in which the drug laws themselves cause individual and group-based harms. This approach focuses on rights discourse. Julie Stewart, the President and Founder of Families Against Mandatory Minimums, in speaking on Lifer Laws, inconsistencies, and the arbitrariness of minimum sentencing, framed this approach thusly:

> I am . . . glad to be on a panel that is about the impact of drug policy on human rights, because sentencing often isn't considered a human rights issue. And yet, I would argue that, in fact, it is, because it is a right to be sentenced fairly when we are convicted of a crime and sentenced to the appropriate sentence given our role in that offense. That is no longer happening now that mandatory sentences are in effect for drug offenses and the handful of other crimes to which they are applied (quoted in Fish, 2000: 83-84; see also Reuter, 1997; Scotti, 2000; Paltrow, 2001; and Boyd 2002 for discussions of the consequences of prohibitionist/punative policies on human/civil rights, and Miron and Zweibel, 1995 for an economic analysis that favors a free market in drugs).

Much drug use HR discourse in the legal arena also acknowledges the historical problem of harms to targeted populations as they are linked with drug use,

> The destructive side of drug antagonism showed itself during the decline phase of the first drug epidemic. Certain groups within American society were feared by the majority. For instance, fueled by racial prejudice, an almost exclusive connection between blacks in the South and the use of cocaine was easily accepted in the first decade of the twentieth century. The turn of the century marked both a peak in the loss of voting rights and of violence towards blacks, as well as a growing fear of violence from cocaine use. Some blacks used co-

caine, as did whites, but an exaggerated linkage developed in the popular mind between cocaine, blacks, and violence, which seemed to justify repression of both the drug and the group. . . . there is a high likelihood that whole groups will be stereotyped by the use of a specific drug. I have mentioned cocaine and blacks before World War I. Chinese were identified with smoking opium, and in the 1930's, Mexican immigrants were linked to marijuana. Also, alcohol Prohibition was in a significant way directed at Catholics and southern and eastern European immigrants. Stereotypes shift over time. I have always found this very interesting because every so often you read that a certain drug is the "natural drug" for an ethnic group. In 1910, heroin was considered a typical "white drug" because it was said to relax competitors in the great social mobility enjoyed by whites. Cocaine was considered a "black drug" because it was said blacks needed a stimulus to do things. But then, in the 1970s, cocaine was thought to be a "white drug" because it provided the edge to win in a competitive life, while heroin was now called a "black drug" because it blotted out the reality of poverty and the ghetto. The powerful symbolic images society absorbs about drugs can make it difficult to deal with them realistically. (Musto, quoted in Fish, 2002: 29-30; 34; see also Tonry, 1994; Reuter, 1997; Steiner and Argothy, 2001; Nunn, 2002;)

Such a historically contextualized reading of approaches to drug use and specific groups within society exposes moral (boundary) reasoning extraordinaire.

The Organizational Landscape Online and Vis-à-Vis Offline Law and Politics

HR as an organizational, activist, and research approach has proliferated and to a large extent become fairly legitimate, at least in terms of announcing discursive positions if not in policy-making and implementation (this latter is for more widespread in some European nations than in the U.S.—see for example Van Vliet, 1990; Tonry, 1994; Reuter, 1997; MacCoun and Reuter, 2002; Lenke and Olsson, 2002; Bergeron and Kopp; Kay, 2002; Erickson, Hathaway, and Urquhart, 2004). Despite this lack of success on the national scene in the United States, because HR has also proliferated in both informal and formal ways across the terrain of the Internet, persons—youth included—located geographically offline in this country have access to a wealth of highly articulate and seemingly legitimate HR resources.

Organizations/Journals with Online Presence

~DanceSafe
~Erowid
~Harm Reduction Coalition: http://www.harmreduction.org/

~*The Harm Reduction Journal*: http://www.harmreductionjournal.com/
home/—overlap with Internet-as-movement—open access to all articles; users defining successful outcomes (Ruefli and Rogers).

~International Harm Reduction Association: http://www.ihra.net/

~International Journal of Drug Policy: http://www.journals.elsevier-
health.com/periodicals/drupol/home

~Canadian Harm Reduction Network: http://canadianharmreduction.
com/

~Centre for Harm Reduction (Australia): http://www.chr.asn.au/home

~International Harm Reduction Development Program: http://www.so-
ros. rg/initiatives/ihrd

~Harm Reduction Project: http://www.ihrproject.org/

~Asian Harm Reduction Network: http://ahrn.thaiis.us/

~HabitSmart's Push Harm Reduction site; billed as first HR info center
on WWW: http://www.habitsmart.com/hrmtitle.html (Site created
by Dr. Robert Westermeyer, Ph.D.—clinical psychologist; addiction and depression specialist).

~The UK Harm Reduction Alliance: http://www.ukhra.org/ (International HR Conference, Belfast, Ireland, 2005)

~Atlanta Harm Reduction Center: http://www.atlantaharmreduction.
org/

~New York Harm Reduction Educators: http://www.nyhre.org/

~Harm Reduction Therapy Center: http://www.harmreductiontherapy.
org/

~Exchange Supplies, Tools for Harm Reduction: http://www.exchange
supplies .org/ (UK)

~Canadian Centre on Substance Abuse: http://www.ccsa.ca/index.asp

~Central and Eastern European Harm Reduction Network: http://www.
ceehrn.lt/

~Drug Policy Alliance: http://www.drugpolicy.org/reducingharm/

~DanceSafe: http://www.dancesafe.org/—offers free webspace under
domain name of www.harmreduction.net

~The New Mexico AIDS InfoNet: http://www.aidsinfonet.org/articles.
php?articleID=155

~Hepatitis C Harm Reduction Project: http://www.hepcproject.org/

~Project on Harm Reduction in the Health Care System: http://www.
temple.edu/lawschool/aidspolicy/default.htm (Scott Burris)

~The Media Awareness Project: http://www.mapinc.org/hr.htm
(Threaded Board for HR)

~CannabisCulture (Canadian—New Democratic Party): http://www.
cannabis culture. com/whatsnew/hr.html

Among these online resources is Students for Sensible Drug Policy (SSDP), an
organization explicitly targeted towards and largely staffed by people in the latter stage of that amorphous group known as youth. Focusing upon this organiza-

tion allows us to get at the question, "What happens when the youth can organize, and talk back in a medium often apprehended as leveling the field and creating authority across at least a few levels of meaning?"

Organizational Case Study

Students for Sensible Drug Policy is an example of a HR organization that exemplifies the youth-adult/community member-citizen transition. Their official mission statement connects their stance on drugs to the larger political process: "Students for Sensible Drug Policy is committed to providing education on harms caused by the War on Drugs, working to involve youth in the political process, and promoting an open, honest, and rational discussion of alternative solutions to our nation's drug problems" ("Mission," 2005). That they are speaking to, and representatives of, a marginal yet potentially powerful group is clear from on the one hand their nascent "national campaign to facilitate dialogue between youth and adults about the harmful impact of prohibition on families," and on the other their voter registration link. Part of their mission is to bring young people into the political process at the local, state, and national levels.

In actual membership, the group claims members that are under the age of majority, but is overwhelmingly peopled by college students. Out of one hundred twenty-four chapters listed on the main website, only two are high school chapters. The chapters are located in all regions of the United States. The Broome Community College of Ithaca, NY chapter website says there are "more than one hundred and fifty college and high school chapters," (http://skateforjustice.org/ media/prelease1.htm) while SSDP-Columbia University says "over two hundred" (http://www.columbia.edu/cu/ssdp/). All of the chapters include contact email addresses for chapter representatives, but only thirty (25%) of the chapters have websites. Of those, fourteen are bare bones, containing a paragraph or less of description of the chapters' activities, history, and status, none containing dated information, thus making ongoing activity levels difficult to gauge. Three sites could not be accessed as of February 4, 2005. A few of the local chapters were not officially recognized by their schools and received no funding as student organizations (such as those at Dartmouth and UW-Eau Claire), but several had on-campus offices and were legitimate and funded student organizations (such as SUNY-New Paltz, UW-Madison, University of Iowa-Iowa City, and Washington University-St. Louis).

While the national organization has four separate campaigns, among the local chapters the campaign receiving the overwhelming amount of attention revolves around the issue that SSDP was founded to address in 1998—repealing the Drug-Free Student Aid amendment to the Higher Education Act. While the amendment remains in place, several local chapters have been successful in getting anti-HEA Amendment resolutions passed in their student senates, and are very active in traditional and online letter-writing and petition drives to their representatives at the national legislative level. One of the most active chapters (at least based on their online representation of their activities), the SUNY-New Paltz group, was successful in changing their school's policy that a first-possession of marijuana would lead to expulsion, to one where the first possession leads instead to a warning.

Diffuse throughout the discourse of the organization(s) online is that of engaging young people in the political process. This is exemplified on the main SSDP site's webpage that offers banners that supporters can add to their websites, where one can find the following bumper-sticker like slogan, "The DARE Generation Speaks Out: End the War on Drugs Youth" [with Youth stylized hand-written over Drugs]. In a local chapter example, the UW-Madison chapter links to an organization, the National Youth Rights Association, that argues for the further lowering of the voting age, citing classic taxation without representation as one of its top ten reasons for doing so. On the page that discusses the Drug Education and Counseling campaign, the SSDP main site states,

> Our organization represents the young men and women who went through drug education and counseling programs that asked abstinence of students. D.A.R.E. is perhaps the most prevalent program. D.A.R.E. targets those students aged ten to thirteen, and police act as the teachers in the program. D.A.R.E. attempts to teach the consequences of drug use, with a heavy emphasis on substances such as marijuana, ecstasy, cocaine, and heroin. One of the disappointments of D.A.R.E. is its participants do not learn the skills and tools essential to making mature, responsible decisions. Rather, through erroneous, extreme examples of drug use and drug users, as well as a stress on the criminal consequences of drug use, police simply attempt to scare students into abstinence. . . . D.A.R.E. is ineffective and unethical for an array of reasons. First and foremost, D.A.R.E. does not achieve its aims of reducing drug abuse and addiction among youth. A federally funded study found that those who are D.A.R.E. graduates are no less likely to use drugs than those who were not a part of the program. Moreover, studies show that D.A.R.E. has a boomerang effect, that some students become interested in drugs due to the D.A.R.E. program. Additionally, D.A.R.E. does nothing for those who do decide to try drugs. Indeed, since D.A.R.E. stresses that all drug use is equally risky and dangerous, some students who use marijuana may think the drug is no more risky or dangerous than cocaine or heroin. Lastly, D.A.R.E. as a program is costly, both in human and monetary resources. We support drug education and counseling programs that are founded on scientific and factual information. Students should be taught the difference between drug use, abuse, and addiction. Students should learn how to lessen the harms associated with drinking and drug use, as well as learn how to avoid an unhealthy, harmful relationship with such substances. Students should feel that all questions are appropriate, and that their answers are accurate. Students should leave a program thinking they truly can resist pressure from peers and outside forces, that they are able to make a mature, responsible decision regarding drinking and drug use. Students should feel that their teachers have treated them with respect. Programs ought to value open and honest education, discussion, and counseling. . . . Our organization sees prohibition as an attack on scientific, factual, useful drug education and counseling, as an attack on youth. (http://www.ssdp.org/home/dec_main.htm)

Herein, SSDP positions itself as clearly HR, but also as a group that seeks to reconstruct the place and status of young people in society.

They are even more explicit in their discussion of the Youth Rights and Privacy campaign:

> Another alarming trend in our country's schools is the tendency to turn to the police, rather than teachers, counselors, and parents, when schools suspect drug use, abuse, or addiction. In January of 2004, the principal of a South Carolinian school asked police to storm the school and search the person and property of many students. He suspected drug use in his school. Drug-sniffing dogs and police entered the school, pointing guns at students and searching them and their belongings. Neither drugs nor weapons were found. Yet, the youth who our drug laws and policies are meant to protect found themselves the victims of these laws and policies. They found themselves scared, humiliated, ashamed, and angry. Our organization sees prohibition as an attack on students' rights and privacy, an attack on youth. When schools turn to police to address drug use, abuse, and addiction, that our country's current approach to drug use and drug users is a failure is evident and undeniable. We will continue to speak truth to the harms of prohibition. We will educate our peers, parents, teachers, and legislators until our drug laws and policies respect the rights and privacy of students. (http://www.ssdp.org/home/srp_main.htm)

Generally, across the national organization's discourse as well as that of the local chapters with an online presence, there is a mission to normalize the participation of what would otherwise be a deviant class. For example, in the profile of the University of New Mexico-Albuquerque SSDP chapter president, Gabrielle Guzzardo, she is presented as a young citizen seeking to be at the center of legal interpretation and policy-making,

> Guzzardo is the founder of the undergraduate chapter of UNM SSDP. She is largely credited with bringing drug policy reform to the UNM campus. Before she initiated the SSDP chapter, there were no active undergraduate drug policy reform groups at UNM. She is a junior in the criminology department and will be going to law school after graduation.

SSDP is thus an organization that takes a stance that while speaking directly to youth, also says to adults (parents in particular), that no matter how much we really want our kids to grow up to be just like us and reflect our particular moral values, we also have to accept their individual decision-making capabilities and tendencies. Lying to them isn't the answer. Force-feeding them one and only one point of view isn't the answer—and in a democracy is irresponsible. Teaching them that anything other than your own point of view is evil can set you up for an estrangement if they discover/decide that this isn't true.

As well, although it is an organization actually populated by the educated middle-class (all chapters are school-based) and, from available visual data on membership, also an overwhelmingly white organization, SSDP takes on boundary-breaking discourse, immersing itself across chapters nationally in the HR discourse that targets racism and classism. As one of its resources, SSDP offers a series of flyers that online users can download, print, and distribute in an effort to engage in the political process. SSDP offers one, "Which of the following crimes will cause you to lose your financial aid?"—certainly a concern of the organization's main constituency of college students—only college students of certain means are going to be directly concerned with such a threat. They also offer a flyer, "Help End the Racist War on Drugs!" directly reaching out beyond the main constituency, and one, "Top 10 Reasons to End the Drug War," which mixes class, racial, and general civil liberties concerns at once. Reason #10 on this flyer directly attempts to create a sense of empathy and solidarity, "There were over 730,000 Marijuana arrests last year, they were people just like you." Not deviants or criminals, but normal folks, are being harmed.

Moral Panic Discourse in SSDP and HR Generally

That SSDP is an organization that is immersed in HR discourse is clear. But an HR stance is not immune from moral panic even though it seeks pragmatic, reasoned solutions to harms, and open discourse about what harm even is. For example, Webster and Stoker both present themselves online as experts in the arena of drug use and treatment. Webster's epistle details his background as an alcoholic member-activist of the Minnesota Model Network (MMN)[8] and participant in the Harm Reduction Coalition's Conferences over a number of years. Stoker's presentation on the history of HR comes to us from his appearance at the World Conference on Drugs. Stoker's expertise is activist-based as well; although he is the Director of the National Drug Prevention Alliance in the UK, he notes, "I can apply my experience to analyzing this situation, and much of that experience, until I moved into the drugs field fifteen years ago, was as a construction engineer" (2001).[9] Both Webster and Stoker use the discourse of crisis, of moral panic, to defend their own personal and organizational stances on opposite sides of the drug use/HR issue.

Webster's take on HR places it on the sacred side of the cultural dichotomy, arguing for its location in medical authority as well as placing substances alongside both that which is most clearly understood as risky, and that which is clearly understood as mundane and normal; "A bold group of caregivers acknowledged that people will engage in potentially harmful behavior in order to obtain a real or perceived benefit. Examples of potentially harmful behavior include substance abuse, bungee jumping and crossing the street. Each behavior has a real or perceived benefit. Each behavior is risky."

In contrast, in his highly critical assessment of HR itself, Stoker locates HR in the wrong-headed permissiveness of the 1970s, and thus places HR on the profane side of the moral dichotomy,

Harm Reduction has always been around. In the Garden of Eden, when Eve ignored the advice to "Just Say No to Snakes" and then peer-pressured Adam into biting that apple, it dawned on them that they were naked and they cried out "What shall we do?" Well, Wal-marts hadn't been invented at that time, so the best they could come up with by way of Harm Reduction was a fig-leaf. And ever since then, we have been using the "fig-leaf" approach to society's drug problems. (2001)

Stoker's discussion of HR clearly is one that casts its opponents as those who misinterpret and misappropriate authoritative cultural arguments (among others in his case, John Stuart Mill's *On Liberty*).

At SSDP itself, the stance that the war on drugs is part of a larger movement in society to disenfranchise young people and illegitimately strip us all of our civil rights/liberties may be understood in moral panic terms as well. In presenting news on a recent Drug Czar press conference on the results of the study "Monitoring the Future" (a dubious choice by the agency in phrasing if the intent is *not* to alarm folks in the HR movement), Tom Angell, SSDP Communications Director, noted the lack of open debate,

Another pointed question was about needle exchange programs (NEPs). Nora Volkow from NIDA stated that NEPs "in the context of drug abuse treatment and HIV education can be an adjunct for reducing HIV." But Walters got up and added that he thought the issue of NEPs is too "divisive" and that we should focus on what we all agree works—treatment. So much for an open and honest discussion about solutions to our nation's drug problems, eh? (www.ssdp.org/ phpBB2/viewtopic.php?t=76)

In the press release for SSDP's response to Drug Czar/Monitoring the Future press conference, Susan Swerdlow, SSDP's executive director, charged Czar John Walters with "distort[ing] reality and ignor[ing] science to suit his ideological agenda" (http://ssdp.org/SSDP_ROOT/5_Press_Releases/Archive/ Press_2005/mtf_release.pdf). Angell prepared and distributed a flyer for the press conference entitled "Drug Czar vs. Reality" (http://ssdp.org/ SSDP_ROOT/5_Press_Releases/Archive/Press_2005/ mtf_handout.pdf), pitting what Walters has said in the media against what the Monitoring the Future research argues, and Walters arguably comes off the worse. [10]

Local chapters of SSDP also publicize obstructionist tactics and response from the government:

The UNM chapter of SSDP has tried twice to set up debates with the Drug Enforcement Administration. Both times, the agency has reneged at the last minute. During the last attempt, the DEA would only agree to participate under strict conditions that greatly limited any challenges to the agency's claims. (see News) We feel that drug policy enforcement organizations are accountable to the public they serve. This means the organizations should be required to engage in open public discourse on drug policy issues. The Drug Enforcement Administration got a case of cold feet the day before a debate with UNM

SSDP. Our guy, Howard Wooldridge, a former police officer and board member for Law Enforcement Against Prohibition (LEAP), was all set to debate the DEA's own Finn Selander. The DEA agent had to back out when his bosses over at the DEA told him that a debate with SSDP was a no-no. It turns out that the DEA has a rule forbidding public debates. Makes you wonder what they have to hide. The UNM chapter of SSDP bent over backwards to make the event happen anyway. The DEA agreed to help, but with a few conditions. The agency was nice enough to send Agent Paul Stone in Selander's place, but there was to be no debate. The two sides would make presentations and then take questions from the audience. Furthermore, the DEA demanded that no media be present at the event. The DEA has the nerve to trample on democracy, execute cruel and unusual drug policies, and act like they know better than doctors what medicines are good for us, But ask DEA agents to justify their actions in a public debate, and they run off with their tail between their legs. Can this organization get any more pathetic?

Thus, as I highlight elsewhere, it is discursively easy to make a reverse moral panic argument that highlights a concern with violating the moral boundary of liberty (Gatson, chapter 9, this volume). However, even with legitimate worries over the state of civil rights/liberties in the wake of The PATRIOT ACTs—especially in a society where *pre*—and post-9/11 activists could make believable arguments that the government was disproportionately and illegitimately surveilling political groups administratively-defined as unpopular or marginal—SSDP isn't getting shut down, or even intimidated, as far as my research has been able to tell. How high its profile even is, is questionable—I didn't even know there was a chapter here at the university in which I work. That, of course, probably says more about my contact with the everyday activities of undergraduates than the profile of SSDP.

Conclusion—Discursive Gamesmanship and Political Publics

The mission of this research project was to analyze the overlap between two sets of emergent behaviors by youth in the context of the rave scene—the use of so-called club drugs in particular and the use of the Internet to seek information about, and perhaps facilitate the use of, those drugs. Although I have suggested that both of these behaviors have been the target of separate and overlapping moral panics, Internet use has also been understood as a valuable surveillance tool—while we may be concerned with the ways in which youth are using the Internet, at the same time, we can be comforted in the ways in which this medium can make just what youth may be doing visible.

We set out to understand the methods and settings of drug use. One finding that we think is important is that the use of CMC and Internet technology more generally highlight that the online world is not one that is disengaged from the offline. However, not only may young people learn dangerous and risky information that they can apply to behaviors and situations in their everyday lives, they may also learn useful, accurate, and powerful information. Indeed, they may learn and take up a serious commitment to political participation. While that need not be seen as neutral and non-risky in and of itself—e.g., one can certainly hold the position that it is undesirable that one's children, or youth in a larger sense, get involved with the "wrong" kind of people, the "wrong" kind of politics—it is a scenario actually intertwined with drug-using and deviant subcultures more generally. These are not actually divorced from politics, however marginalized their populations may be. Defining marginalized people is itself a set of political acts, and those very people may be more or less politically responsive.

Further, we find a reiteration of practices that are merging the online with the offline, and support the conclusion that there is in actuality no virtual world (see Gatson and Zweerink, 2004: 14; 39-41; 141; see also Tuszynski, 2004; Rodman, 2003: 26-27). Real information, contact with real persons, and often using online spaces to express that which is not always safe to express offline are typical ways in which youth are making use of the Internet in a drug-aware (not always a drug-*using*) subculture. Again, whether or not one sees this as a positive, the genie is pretty much out of the bottle—will we also see this as a Pandora's box of issues, behaviors, and problems which we wish we could stuff back inside that box?

Political Compromise—Where Does the Middle Ground Lie?

Gilbert Rodman makes an arguably important distinction between democratic participation and participatory democracy (2003: 28-32). In cautioning against a utopian reading of the possibilities for power-sharing in Internet forums, he argues,

> this power appears more revolutionary than it actually is precisely because of the ways that other media have taken over (and eviscerated) the public sphere: in a society where the average citizen rarely has access to a public forum where he or she can *share* (and not just consume) opinions and ideas, the ability to "publish" one's thoughts where potentially millions of people might read them is a dramatic deviation from the status quo. At the same time, the extent to which this power actually makes the Net a democratic space is questionable, and we need to be cautious about conflating the power that individual users have to "speak" online with actual power over the networks that comprise the Net. (2003: 29-30)

Additionally, Rodman asserts that issues of actual access to such power-sharing venues have not even begun to be overcome (2003: 32-38). Rodman emphasizes the messiness of the Internet, in both experiences and the polysemics of those experiences (2003: 39). I think we should take that messiness itself as a serious thing—shouldn't democracy be somewhat unstable, unsettled? Shouldn't our debates always be in a state of unstable equilibrium (see Omi and Winant, 1994: 84-88)?

Ann Travers asserts, "behaviours of both inclusion and exclusion are learned in community," and urges educating in the direction of "learn[ing] to behave in ways that encourage rather than discourage broad participation" (2000: 132). One thrust of her recommended practices includes the formation of codes of conduct from within situated online communities, emphasizing the clashes between the hegemonic understandings of freedom: freedom from versus freedom to. Travers argues that you can establish *freedom from* without shutting down critical discussion (2000: 139-140).

Travers's Critical Dialogue "assumes a multiplicity of perspectives and is embedded in an attempt to construct community and establish inclusive public space" (2000: 132). She is thus focused on an idea of harm reduction to the body politic. Arguing from a specifically feminist perspective, she introduces the concept of counterpublics, noting the importance of such subaltern spaces for re-grouping and re-confronting the larger society (Travers, 2000: 147; see also Omi and Winant's discussion of Gramsci's concept of war of maneuver, 1994: 81). When these counterpublics are really public—i.e., accessible—beyond the counterpublic's constituency, the possibilities for critical discussion and deliberative speech, particularly on moral subjects, may be truncated.

However, we must take outsiders seriously, and not because "we" fear them, and they are risking "our" society. Recognizing that Pogo's classic line— "We have met the enemy and he is us"—is meaningful in a networked (offline and online) society is, I think, crucial.[11] We must take Jed Miller's notion of developing deliberative spaces, and particularly deliberative spaces for children and youth, seriously (2004). It is actually easier, cheaper, and more sustainable to do this online, and it is also, I would argue, safer, if what we are deliberating includes risky behavior. No moral panic campaign has ever succeeded in stamping out whatever its targeted behavior(s) have been. We would do well to keep this in mind, and if harm reduction is no panacea, surely deliberative democracy must be.

Notes

1. In a case that demonstrates the similarity between youth violence and moral panics about mass media, Barker discusses the case of Alan Poole, a young man whose crimes led to a shoot-out with police in which he also died; his case was discussed in news, political, and scholarly accounts of the 1950s as an exemplar of youth violence. Eerily akin to the discussions of the causal influences of the young gunmen in the Columbine shootings, "'for every child who, identifying himself with some gunman hero, develops actual criminal tendencies, there must be many thousands upon whom the constant reading of these comics must have a deeply unwholesome effect.'" It was noted that Poole "'had a library of fifty of these [horror] comics. Indeed, one social worker said that he had a collection of three hundred . . . we found that this particular lad had one "Western" comic in his possession, and that not a very alarming one'" (1984, 30). Musto also

notes such exaggerations and outright lies in crusades against drugs, "Hobson . . . declared in the mid-1920s that one ingestion of heroin, even on an ice cream cone, could addict a child" (quoted in Fish, 2000: 32).

2. Or as Musto argues, "Jerry Rubin was said to have experimented widely with drugs in the 1960s. When I met him in the 1990s, he not only wore a jacket and tie, but also had a dietitian coming twice a week to his apartment to go over his diet for that week so that he would eat the healthiest food possible. Jerry represents the great shift from the attitude of the 1960s and 1970s to one of caution and risk reduction that involves many more aspects of life than just drugs" (quoted in Fish, 2000: 25).

3. Neither, as Fish points out in his summary of the 2000 Conference in New York on the effectiveness of and alternatives to U.S. drug policy, do "the penalties for the various licit and illicit substances bear [any] relationship to their dangerousness on these two critically important dimensions [of "safety margin" and "dependence potential"] (2000: 13).

4. However, I will discuss below the ways in which HR itself does take on moral panic discourse.

5. Harcourt also, in my reading of the literature both scholarly and from the HR movement, errs in asserting that the harm principle is obsolete, "collapsing under the weight of its own success" and that "Today, the issue is no longer whether a moral offense causes harm, but rather what type and what amount of harms the challenged conduct causes, and how the harms compare. On those issues, the harm principle is silent." (1999: 113). True, any reading of even only the drug use HR discourse shows these proponents clearly compare the harm of drug use to the harm of incarceration, stigma, and unequally applied drug laws, as well as comparing the harm to pain patients of denying them access to drugs with the harm to society of recreational, private, small-amount drug users. As well, the abstinence/prohibition, or anti-HR, discourse clearly compares the harms to society and youth (in particular) with any harm done to users and sellers of drugs—the latter is not harm, but just punishment. This indeed may not be the harm principle as Mill defined it, but outside of Harcourt's narrow reading of the harm principle only in "legal philosophy and rhetoric"—in a larger, more diffuse sociocultural discourse—I think we can see the harm principle in effect (1999: 118). Mightn't it be that Mill was not explicit in comparing harms because he was offering an abstract principle that was meant to be applied to particular situations? In saying, "the fact of living in society renders it indispensable that each should be bound to observe a certain line of conduct toward the rest," Mill is rather imprecise as to where that line is (quoted in Harcourt, 1999: 121). Harcourt acknowledges both the abstract, polysemic quality of Mill's work, as well as his later precise statements regarding regulation of particular harmful activities (1999: 121-22)—in deciding which things to regulate, Mill himself must have compared harms; what he considered harmful must be understood culturally, and how any of us chooses to use Mill must also be so considered (see for example Stoker, 2001, in contrast to Harcourt).

6. 149 other projects were targeted in an eleven-page list, and many of the research projects involved the study of human sexuality (Yeoman, 2004).

7. In some online publications, NIDA uses the phrase "drugs of abuse," and while advocating that youth especially never *start* using drugs, also asserts, "We now have scientific evidence showing that people can't always choose to stop using drugs" (www.nida.nih.gov/drugpages/PSAs.html).

8. Webster describes the Minnesota Model thusly, "a group at Hazelden in Minneapolis usurped the Twelve Steps of AA and repackaged them as the Minnesota Model" (2005).

9. Webster's page that contains his email is actually one of several places to find a link to Stoker's discussion.

10. See also Barker for a similar reverse moral panic approach. In his argument there *was* a danger in comics, but not the one the campaign against them asserted. Barker instead focuses on the subgenre of Korean War comics, which "contained a virulent anti-Communism, which verged on racism" (1984: 200). But Barker argues that concern over gratuitous violence was the misplaced panic, "Wouldn't it be far more effective to present an innocent, unalienating picture *through which* readers might be drawn, unsuspecting, into collusion with the selling ideology?" what he calls a "normalization of America" (1984: 201).

11. "There is no need to sally forth, for it remains true that those things which make us human are, curiously enough, always close at hand. Resolve then, that on this very ground, with small flags waving and tinny blast on tiny trumpets, we shall meet the enemy, and not only may he be ours, he may be us." (Kelly, 1952)

References

Cases Cited
Abrams v. United States, 250 U.S. 616 (1919).
Schenck v. United States, 249 U.S. 47 (1919).

American Heritage Dictionary. Boston, MA: Houghton Mifflin Co., 1991.
Barker, Martin *A Haunt of Fears: The Strange History of the British Horror Comics Campaign*. Jackson: University Press of Mississippi, (1984).
Becker, Howard S. *The Outsiders*. New York, NY: Free Press, 1963.
Beisel, Nicola. *Imperiled Innocents: Anthony Comstock and Family Reproduction in Victorian America*. Princeton Studies in American Politics: Historical, International, and Comparative Perspectives Series; editors Ira Katznelson, Martin Shefter, and Theda Skocpol. Princeton, NJ: Princeton University Press, 1997.
Bergeron, Henri, and Pierre Kopp. "Policy Paradigms, Ideas, and Interests: The Case of the French Public Health Policy toward Drug Abuse." *Annals of the American Academy of Political and Social Science*. Vol. 582 (2002): 37.
Blumenson, Eric. "Recovering From Drugs and The Drug War: An Achievable Public Health Alternative." *Journal of Gender, Race and Justice*. Vol. 6 (2002): 225.
Bowman, Frank O., III. "The Geology of Drug Policy in 2002." *Federal Sentencing Reporter*. Vol. 14 (2002): 123.
Boyd, Graham. "Collateral Damage in the War on Drugs." *Villanova Law Review*. Vol. 47 (2002): 839.
Brown, Daryl K. "Rationing Criminal Defense Entitlements: An Argument from Institutional Design." *Columbia Law Review*. Vol. 104 (2004): 801.
Clemens, Elisabeth. Review of *Imperiled Innocents: Anthony Comstock and Family Reproduction in Victorian America*. *American Journal of Sociology*, 103, no. 5 (1998): 1483-1485.
Cooper, Elizabeth B. "Social Risk and the Transformation of Public Health Law: Lessons From the Plague Years." *Iowa Law Review*. Vol. 86 (2001): 869.
Edles, Laura Desfor. *Cultural Sociology in Practice*. Malden, MA: Blackwell Publishers, 2002.

Erickson, Patricia G., Andrew D. Hathaway and Cristine D. Urquhart. "Backing into Cannabis Reform: The CDSA and Toronto's Diversion Experiment." *Windsor Review of Legal and Social Issues*. Vol. 17 (2004): 9.

Fire Erowid. "Face to Face with NIDA: A Conference on Drugs, Youth, and the Internet." *Erowid Extracts: A Psychoactive Plants and Chemicals Newsletter*. Number 3 (2002): 2.

Fish, Jefferson M. "Conference: Is Our Drug Policy Effective? Are there Alternatives?" *Fordham Urban Law Journal*. Vol. 28 (2000): 3.

Gatson, Sarah N. "When Do Young People Become Community Members and/or Citizens?" Summer Institute on Digital Empowerment: The Internet and Democracy, Center for Digital Literacy, Syracuse University, Syracuse, NY, July 8-9 (via Telecast). 2004.

Gilmore, Norbert. "Drug Use and Human Rights: Privacy, Vulnerability, Disability, and Human Rights Infringements. *Journal of Contemporary Health Law and Policy*. Vol 12 (1996): 355.

Goldkamp, John S. "The Drug Court Response: Issues and Implications for Justice Change." *Albany Law Review*. Vol. 63 (2000): 923.

Henningfield, Jack E., and John Slade. "Tobacco-Dependence Medications: Public Health and Regulatory Issues." *Food and Drug Law Journal*. Vol. 53 (1998): 75.

Heyman, Steven J. "Law and Cultural Conflict: Ideological Conflict and the First Amendment." *ITT-Chicago-Kent Law Review* 78 (2003): 531.

Hunter, Nan D. "Living with *Lawrence*." *Minnesota Law Review*. Vol. 88 (2004): 1103.

Inciardi, James A., and Lana D. Harrison, editors. *Harm Reduction: National and International Perspectives*. New York, NY: Sage Publications, 1999.

Jensen, Eric L., and Jurg Gerber, editors. *The New War on Drugs: Symbolic Politics and Criminal Justice Policy*. Cincinnatim, OH: Anderson, 1998

Kay, Amanda. "The Agony of Ecstasy: Reconsidering the Punitive Approach to United States Drug Policy." *Fordham Urban Law Journal*. Vol. 29 (2002): 2133.

Kelly, Walt. *The Pogo Papers*. New York, NY: Simon and Schuster, 1952.

LeBlanc, Lauraine. *Pretty in Punk: Girls' Gender Resistance in a Boys' Subculture*. New Brunswick, NJ: Rutgers University Press, 1999 (2000).

Lenke, Leif, and Boerje Olsson. "Swedish Drug Policy in the Twenty-First Century: A Policy Model Going Astray." *The Annals of the American Academy of Political and Social Science*. Vol. 582 (2002): 64.

Lindesmith, Alfred. *The Addict and the Law*. Bloomington, IN: Indiana University Press, 1965.

MacCoun, Robert. "Toward a Psychology of Harm Reduction." *American Psychologist* 53 (1998):1199-1208.

MacCoun, Robert, and Peter Reuter. "The Varieties of Drug Control at the Dawn of the Twenty-First Century." *The Annals of the American Academy of Political and Social Science*. Vol. 582 (2002): 7.

Malkin, Ian. "Establishing Supervised Injecting Facilities: A Responsible Way to Help Minimise Harm." *Melbourne University Law Review*. Vol. 25 (2001): 680.

Miller, Jed. "Toward an Interactive Democracy: Prospects and Challenges for Fostering a Deliberative Public Sphere Online," paper for The Kettering Foundation. 2004 <http://jedmiller.com> (7 Jul. 2004).

Miron, Jeffrey A., and Jeffrey Zwiebel. "The Economic Case Against Drug Prohibition." *The Journal of Economic Perspectives* 9, no. 4 (1995): 175-192.

Nunn, Kenneth B. "Race, Crime and the Pool of Surplus Criminality: Or Why the 'War on Drugs' was a 'War on Blacks.'" *Journal of Gender, Race, and Justice*. Vol. 6 (2001): 381.

Omi, Michael, and Howard Winant. *Racial Formation in the United States: From the 1960s to the 1990s*. New York, NY and London: Routledge, 1994.

Paltrow, Lynn M. "Pregnant Drug Users, Fetal Persons, and the Threat to *Roe v. Wade*." *Albany Law Review* 62 (1999): 999.

Rasmussen, David W., and Bruce L. Benson. "Rationalizing Drug Policy under Federalism." *Florida State University Law Review*. Vol. 30 (2003): 679.

Reuter, Peter. "Why Can't We Make Prohibition Work Better?: Consequences of Ignoring the Unattractive." *Proceedings of the American Philosophical Society*. 141, no. 3 (1997): 262-275.

Rodman, Gilbert B. "The net effect: The public's fear and the Public Sphere." Pp. 9-48 in *Virtual Publics: Policy and community in an electronic age*, edited by Beth E. Kolko. New York: Columbia University Press, 2003.

Rohall, David, editor. "From the polls: Illegal Drugs." *Contexts*, 3, no. 2 (2004): 60.

Scotti, Roseanne. "The 'Almost Overwhelming Temptation': The Hegemony of Drug War Discourse in Recent Federal Court Decisions Involving Fourth Amendment Rights." *Temple Political and Civil Rights Law Review*. Vol. 10 (2000): 139.

Spade, Dean. "Undeserving Addicts: SSI/SSD and the Penalties of Poverty." *The Social Justice Law Review*. Vol. 5 (2002): 89.

Steiner, Benjamin D., and Victor Argothy. "White Addiction: Racial Inequality, Racial Ideology, and the War on Drugs." *Temple Political and Civil Rights Law Review*. Vol. 10 (2001): 443.

Stoker, Peter. "The History of Harm Reduction." Paper given at the World Conference on Drugs, Visby, Sweden—May 3rd to 6th, 2001. <http://www.dpna.org/resources/positions/harmreduction.htm> (21 Jan. 2005).

Tonry, Michael. "Race and the War on Drugs." *University of Chicago Legal Forum*, 1994.

Travers, Ann. *Writing the Public in Cyberspace: Redefining Inclusion on the Net*. Garland Studies in American Popular History and Culture, edited by Jerome Nadelhaft. New York and London: Garland Publishing, Inc, 2000.

Tuszynski, Stephanie. "Response to Marcy Chvasta's "Screening Bodies: Performance and Technology." In *Performance/Text/Technology*. 2004. <http://www.cyberdiva.org/PTT/responses/stevie.html> (May 2004).

van Vliet, Henk Jan. "The Uneasy Decriminalization: A Perspective on Dutch Drug Policy." *Hofstra Law Review*. Vol. 18 (1990): 717.

The Vaults of Erowid: Documenting the Complex Relationship Between Humans and Psychoactives. <http://www.erowid.org/> (2002-2003).

Webster, Peter. "The Victims of Prohibition—2004 Harm Reduction Conference." Posted email to James Cannon, January 3. 2005. <http://mail.psychedelic-library.org/show.cfm?postid=7045androw=6> (21 Jan. 2005).

Yeoman, Barry. "Forbidden Science." *Discover* 28, no. 8, 2004.

Chapter Three

The New Drugs Internet Survey: A Portrait of Respondents

Edward Murguía and Melissa Tackett-Gibson

Our study began as an exploration into the relationship between club drug culture, raves, and online discourse. The goal was to solicit new and pertinent information related to the context and setting of club drug use. For example, we wanted to simply document when, where, and with whom drugs are primarily used; under what circumstances are some drugs chosen over others, and, most importantly, we wanted to explore the role of the Internet in transferring and mediating information regarding drug use. Therefore, the primarily purpose of the study was to provide an exploration not only of patterns of drug use among "ravers," but the role technology plays in creating and disseminating drug use trends.

We came to realize that we had launched into the world of online Harm Reduction communities. These communities were large, complex, and varied in their specific missions. One site emphasized drug testing and pill identification; another focused on providing access to a plethora of medical studies and other information on specific substances. While several websites had grown up out of the club drug/rave movement, at the time of the study they had little to do with raves per se and were far more involved in drug education and cultural expression. Interviews suggested that several of the major sites shared rivalries in terms of the numbers of members, forums, and discussion threads (DanZerD, interview with author, July 9, 2002). One site was frequently described online as

45

"cutting edge" in terms of their drug education content; another was often described as "reputable" in online discussions. Clearly, the online world of drug information was large, diverse and challenging in terms of study.

With little information on such communities from previous sources, we adopted multiple methods to begin to characterize those groups and individuals participating in harm reduction and/or drug information discourse online. The bulk of our work, including the articles in this volume, relied upon qualitative assessments of site content and discussion. To this we wished to add quantitative data that might provide information representative of the larger harm reduction community online. The following sections describe the process by which we recruited website communities to survey participation, an overview of the surveys content, and the specific methods of survey administration. Lastly, we present a descriptive overview of survey responses.

Survey and Study Recruitment

In 2002, we identified five major sites to include in the study, three of which maintained large discussion forums. Each site was contacted and asked to assist us in the development and administration of the projects survey. We also asked for permission from site administrators to participate in and observe site discourse. Of the five contacted, two indicated interest in study participation. Each obtained from us privacy agreements and an overview of study objectives and methods. Ultimately, one site expressed concerns over member privacy and declined to participate. DrugSite (a pseudonym) remained in the study and its administrators agreed to revise and comment on survey content; pre-test the survey and allow researchers to anonymously read (i.e., lurk) at posting sessions with public access and in private computer-mediated communications (CMC) areas. They also agreed to promote survey participation among site members. This included an announcement about the survey on several major discussion boards, and a link to the survey on DrugSite's home page. At that time, site administrators indicated that theirs was the largest online harm reduction community on the Internet with over thirty-five thousand members.

Survey Development and Administration

The survey was developed as an online instrument. It included a total of 194 questions with skip patterns. There were six open-ended questions. The questions fell into categories including:

- the type and speed of Internet connection;
- the level of Internet use;
- online and face-to-face relationships with friends;
- sources of drug information;

- use and perceptions of drug-related websites;
- impact of online drug information;
- rave/party attendance and drug use;
- prescription drug use;
- past thirty days use of various substances;
- emotional distress indicators; and
- demographics.

The survey was reviewed by DrugSite moderators who reviewed our drug terminology for accuracy. Fifteen moderators also pre-tested the online instrument for functionality and provided comment for improvements.

Given the survey's length and because of the nature of its content, we wanted to make the survey as free of constraints and complications as possible. Thus few validation rules (other than those excluding minors from participation) were included in the survey. In addition, no passwords were required. We compensated for this broad access by identifying and excluding potential duplicate surveys post hoc from the dataset. Duplicate surveys were defined as those submitted within three hours of each other with a majority of identical answers.

Participants were recruited to the survey primarily through DrugSite. Solicitations were also forwarded to contacts from other drug-related websites by email. While the others sites did not formally promote the survey, some respondent were obtained through word-of mouth advertising. In total we received 1,038 responses, 82% of which came to the survey as a result of DrugSite's referral. Due to the survey's length, responses were examined to determine roll-off. 87% of the participants completed the last major question of the survey.

Results from the New Drugs Research Group Survey

Demographics: Age, Gender, Residence, Marital Status, Children

Our sample was limited to those eighteen years old or older; in other words, our survey eliminated those less than eighteen years of age for whom we would have been required to receive parental permission before surveying them. Despite this stricture, respondents to the survey tended to be young; ninety percent were under thirty years of age, and 95% were under thirty-five years of age. Only 1% of the respondents were forty-five years old or older. The youth of the respondents reflects the fact that young people have been the quickest to accept the Internet, and may say something as well about the demographics of those who use non-legal drugs. Additionally, a very large percentage (76%) of the respondents, were male.

We assumed that almost all of the respondents to the survey were going to be from the United States, given that the website on which we advertised the survey was popular and heavily used by American youth. The reality was that only 61% of respondents were from North America and, of the North American respondents, only 80% were from the United States. In other words, of the approximately 1,000 respondents (n=1,038), about one-half were from the United States. 23% of the respondents were from Australia/New Zealand, and 14% were from Europe. Of respondents from the United States, 25% were from the Northeastern section of the United States, 23% were from the Southeast, 22% were from the Midwest, 18% were from the Southwest, and 12% were from the Northwest.

Most of the respondents (78%) were single, never married. The second largest category, however, was not married and living with a spouse, as we imagined it would be. The second largest category (11%) was "living with domestic partner," that is, living in a cohabiting relationship. Only 7% of respondents were married and living with their spouse. Again, this reflects the fact that the respondents were a young group, most not having been married, and that among youth, cohabitation is becoming a more acceptable option than it has been among previous generations. The overwhelming number of respondents (93%) did not have children. Again, this reflects the youth of this population. It could also be, though, that having children adds responsibilities militating against the use of drugs.

Race, Schooling, Employment, Residence, and Income

Survey respondents were overwhelmingly (92%) white. Less than 1% of respondents indicated that they were black, 2% indicated that they were Asian, and .03% indicated that they were Pacific Islanders.

Concerning schooling, a majority (57%) of respondents were still in school. Of those in school, 37% were in college full-time, 9% were in high school, and 9% were in college part-time. In other words, of those in school, 46% were in college and 9% were in high school. Certainly, part of this configuration is the result of our request that only those eighteen and over respond to our Internet survey which limited the number of high school students who could respond to the survey.

With reference to the highest degree or level of school completed, 43% of respondents had either completed high school coursework, had received a high school diploma or its G.E.D. equivalent, or had received a vocational certification. Over one half (51%) of the total respondents had some college credit, and 23% had bachelor's degrees, master's degrees, professional degrees, or doctoral degrees. On average, then, respondents tended to be well educated.

Regarding current employment, 41% of respondents were employed full-time, 28% were employed part-time, 13% were unemployed and looking for work, and 15% were not employed and not looking for work, such as students and homemakers. What is striking here is that over 40% were full-time workers. This belies the stereotype that individuals who visit Harm Reduction sites are largely unemployed and/or part-time workers.

At what occupations did respondents report working? The largest percentage of respondents (16%) indicated that they worked in the area of retail management and sales. It could be that the sometimes alienating nature of sales work encouraged these workers to seek drug use. The second largest occupational category was that of computers and technology. It is not unexpected that a large number of individuals in this occupational category would have interest in the Internet and in Internet sites having to do with drugs and Harm Reduction, first because their focus is on information technology itself, and second, because some of the culture of information technology work involves mood enhancing and/or performance enhancing substances. The third largest category (12%) was of individuals in office/administrative support, and, finally, the fourth largest category (11%) was of workers in the food management and service industry. Only 1% of respondents worked for the government or the military. Clearly, given the fact that most of the drug use discussed on the Internet website on which we advertised involved drugs that were not legal, government and/or military workers would have had a great deal to lose were they to engage in the use of non-legal drugs, and this is probably a reason that we received very few responses from them.

What was the living situation of respondents? The greatest number of respondents (42%) lived with parents, guardians, or in laws, 25% lived with friends and roommates, and 19% lived with their spouse/partner. 34% indicated that they lived in a house or apartment owned either by themselves or their parents but with a mortgage. 21% indicated that they lived in a house or apartment owned outright (without a mortgage) by either themselves or their parents. 42% reported living in a house or apartment rented either by themselves or their parents, and only 2% reported living in a house or apartment neither owned nor rented.

Table 1 below indicates the approximate annual household income of respondents, including the annual salaries of everyone living within the household. What is striking about this table is the relative lack of variation among the income categories. For example, the largest income category in terms of % of respondents is from $20,000 to $35,000 U.S. dollars at 15%, while the smallest income category of the eight categories is over $150,000 U.S. dollars at 8%. The relationship, then, between drug use and social class, despite a media emphasis on lower and working class drug use, is not strong.

Table 3.1. Annual Combined Household Income

Income in U.S.D.	Frequency	Percentage	Cumulative Frequency	Cumulative Percentage
Less than $20,000	135	13.9	135	13.9
$20,001 to $35,000	145	14.9	280	28.8
$35,001 to $45,000	121	12.5	401	41.7
$45,001 to $60,000	144	14.8	545	56.1
$60,001 to $80,000	143	14.7	688	70.8
$80,001 to $100,000	106	10.9	794	81.7
$100,001 to $150,000	97	10.0	891	91.7
Over $150,001	81	8.3	972	100.0

Source: DrugSite Internet Survey (2003).
Note: Frequency missing = 66.

Respondents and Computer Usage of Respondents

We asked when respondent first found out about our survey, and the over-whelming number of respondents found out about the survey from the drug use Harm Reduction website on which we advertised. 82% of respondents said that they first found out about the survey from the Harm Reduction website, 5% said that they learned about the survey from a friend, 5% from another website, 4% from us as researchers, and 1% from a search engine.

We then asked how much time respondents spent participating on website discussion boards. If we consider those spending two or more hours on website discussion boards as very active on the Internet, this comprises 23% of respondents who answered this question. In other words, about one quarter of respondents spend a considerable amount of time on website discussion boards. Additionally, over 30% of respondents actively participated on the website discussion boards that they follow most closely. Thus, 35% of respondents indicate that they were participants who post frequently. An additional 47% are participants who did post but not frequently and only 19% were "lurkers" who did not post but only read the boards.

We asked respondents if they had an Internet connection at home. An over-whelming percentage (94%) of those responding to this question said that they did have an Internet connection at home. We then asked respondents how their home computer was connected to the Internet. The percentage of respondents who said that they were connected by means of a cable modem was 40%, while 28% said that they were connected via telephone dial-up modem, and 25% indicated that they were connected by means of DSL (Digital Subscriber Line). These three methods make up 93% of the respondents who answered the question. We asked if respondents had an Internet connection both at work and at school. 41% of respondents who answered this question answered that they did have an Internet connection at work and 55% answered that they did have an Internet connection at school.

We wanted to know where the respondents primarily connected to the Internet. A large majority (82%) indicated that they primarily connected to the Internet at home. Only 11% indicate that they primarily connect to the Internet at work, 4% primarily connected at school, and only 2% said that they primarily connected to the Internet at a public place, such as a cafe or a library. Clearly, then it is at home where respondents felt that they have the privacy to engage on the Internet with drug related issues.

We wanted to know how much time respondents spent sending, receiving, or reading mail on the Internet. Over one half (54%) of respondents who answered this question indicated that they spent thirty minutes or less on the Internet in an average day, about twenty percent (20%) indicated that they spent between thirty minutes to an hour on the Internet, and almost ten percent (10%) said that they spent between one and two hours.

We asked if most of the mail received by our respondents were from mailing lists. Although a considerable amount of mail was received from mailing lists, most respondents report that the larger part of their email was not from these lists. 67% of respondents report that none or less than half of their mail is from mailing lists. In parallel fashion, respondents report that most of the email they send is not to mailing lists. Thus, a total of 94% of the mail respondents send is not to mailing lists.

Topics on the Internet

We asked respondents to select the one category that best describes the types of topics discussed on the email list that they read most often. The category selected most often was the "music, clubs, or parties" category at 31%. Next, they selected the "health, drugs, or safety" category (20%). We should remember that most of the respondents learned about this survey from a drug use Harm Reduction site. Also, to no surprise, the fact that the above two categories were selected most often tells us that music, clubs, parties, and drug use are connected.

Our concern both in this chapter and in this book is with information about drugs and drug use on the Internet, and so we asked, "How often are recreational drugs and/or drug use discussed on the email list that you read MOST often?" If we combine the categories of "sometimes," "often," and "always," we find that 38% of respondents reported that drugs and/or drug use was discussed on the email list that was read most often. Again, this is not surprising because most of our respondents were recruited from a drug use Harm Reduction site.

When asked the one category that best described the types of topics discussed on the web discussion board that the respondent followed most closely, by far the largest response category was the "health, drugs, or safety" category at 61%. The second large response category was the "music, clubs, or parties" category at 17%. Our next question was, "How often are recreational drugs and/or drug use discussed on the web discussion board you follow most closely." Again, not unexpectedly, a total of 88% of respondents indicated that recreational drugs and/or drug use was the web discussion board followed most closely by the respondents.

Real-time Chat, Instant Messaging, and Conversations about Drugs

In our study, we had a particular interest in online community because we believe that the Internet is changing the meaning of "community" as it refers to the use of drugs. Therefore, we asked how much time respondents spend participating in real-time chats, instant messaging, etc. While 22% said none, fully 37% said that they spent up to one hour in real-time activities on the Internet. The percentage of respondents spending over two hours in real-time activities on the Internet was 24%. We then asked respondents to indicate how often they discuss recreational drugs and/or drug use through real-time activities on the Internet.

Only 8% say that they never do. The percentage of those who respond that they often or always discuss recreational drugs and/or drug use was large, namely, 62%. This is one of the most striking findings in our survey, and this indicates that, as we had thought, for some of the population who uses mood enhancing substances, the concept of community relative to drug use is being changed by the Internet, and geographic, physical space is being replaced in part by cyberspace.

We wanted to know about networking on the Internet, and so we inquired as to where respondents first met the individuals with whom they were communicating online. Most were not actually first met online. About two thirds of respondents (69%) of respondents report that they met none or only one or two of the close friends with whom they were communicating online. Only 8% of the respondents met more than one half or all of the close friends with whom they were communicating online.

We asked about the respondents' "best friend" whom they met online, and to what extent respondents, first, communicated with this best friend by email, by telephone, or face-to-face, second, the geographical distance they lived from their friend, and, third, whether or not respondents met their best friend through a drug-related website or email list. About one third of respondents (34%), reported that they corresponded with their best friend met online by email at least weekly or daily. A similar 36% of respondents report that they correspond with their best friend met online at least weekly or daily by telephone, and about one fourth (27%) of respondents report seeing their best friend met online face-to-face at least weekly. Concerning physical proximity to best friend, respondents report that 48% of best friends met online live within fifty miles, but 35% of best friends, according to the respondents, live more than two hundred miles away. This finding is theoretically important, because it demonstrates that the Internet has changed the concept of "community," making members no longer bound by geographical distance for instant communication. Although this has been true in the past with letters and the telephone, the Internet has substantially improved communication with others regardless of physical proximity. Finally, about one third of best friends (33%) whom the respondents met online were met through a drug-related website or email list.

We asked first about respondents' relationships with their closest friend who was initially met face-to-face, that is, a friend met originally at school, work, through other friends, but not online. We found that about three fourths (75%) of respondents report that they discuss recreational drugs and/or drug use with their closest friend initially met face-to-face either sometimes, often, or always. Even though this best friend was initially met face-to-face, one third (35%) of respondents report that they communicate weekly or daily by email with this friend. A larger percentage of respondents (69%), report that they communicate either weekly or daily by telephone with their closest friends originally met face-to-face. This makes sense, since, most probably, friends originally met face-to-face would be within local calling distance and talking on the telephone is to be preferred with communicating by means of email, because on the telephone, one can hear the other's voice and can be attuned to the nuances of what the other person says, unlike with the written word. Also, there is the advantage of real-time communication with instantaneous response when one is talking on the telephone, unlike when one is communicating with email (other than instant messaging). Finally, it is easier to speak than to write one thoughts, as one has to do when one is using email.

Over two thirds (68%) of respondents report that they continue to communicate with their closest friend met face-to-face on a face-to-face basis at least weekly. This 68% rate is much larger than the 27% of respondents who report that they communicate with best friends not met face-to-face, but rather met online on a weekly or daily basis, as one would expect. Also, as expected, respondents' best friends met face-to-face tend to live geographically close to respondents. Respondents report that about two-thirds (68%) of their best friends whom they met face-to-face live within twenty miles, and 78% of these best friends live within fifty miles of them. The 77.6 percentage can be compared to the smaller 48% of best friends met online who live within fifty miles of respondents. Also, whereas 33% of best friends met online were met through a drug-related website or email list, only 14% of best friends met face-to-face were met in a parallel way at a party, dance, or club, presumably where mood altering substances would be available. With best friends met face-to-face, respondents report discussing recreational drugs and/or drug use either sometimes, often, or always 83% of the time. This can be compared with a previous finding where respondents report discussing recreational drugs and/or drug use either sometimes, often, or always 75% of the time with best friends met online. Because the two numbers are so similar, we can infer that it makes little or no difference where one's best friend was first met if a respondent is immersed in a subculture which involves the use of drugs.

We next attempted to ascertain to whom our respondents go when they want information about a recreational drug, such as marijuana, MDMA, mushrooms, cocaine, etc. We asked whether or not respondents asked various individuals for information about recreational drugs. Not unexpectedly, we found that respondents were most likely to talk with friends (73%), coworkers, classmates, or casual acquaintances (38%), and spouse/partner/boyfriend/girlfriend (32%). They were least likely to discuss drugs and drug use with school teachers/school professionals (5%), counselors or mental health professionals (7%), club owners, DJs or bartenders (10%), or parents (10%). Clearly, the greatest source of information about recreational drugs were friends of respondents, and the second most important source were coworkers, classmates, or casual acquaintances of the respondents (which category may overlap with the "friends" category. Authority figures such as teachers, mental health professionals, and parents were the least likely to be consulted.

Respondents were then asked, "In general, what information about recreational drugs were you interested in getting?" There were seven response categories to this question, and the number of respondents giving a positive response to each response category was as follows: how the drug makes you feel (50%);

potential drug interactions (42%); how to use the drug (39%); potential drug side effects (51%); potential long-term side effects (44%); where to obtain the drug (41%); and other information (7%). Because the percentage of respondents seeking "other information" is so low, it is probable that we have asked the major questions concerning information about recreational drugs that people want to know. There is no major variation on the seven categories of information requested, although the two categories with the highest response numbers are information on potential drug side effects, and information on how the drug make one feel.

We then inquired of respondents as to how they themselves asked for information about recreational drugs. There were six response categories to this question, and the numbers of respondents who gave a positive response to each category is as follows: face-to-face (67%); telephone (22%); AIM (AOL Instant Messenger) (17%); email (15%); text messaging (14%); and regular mail (3%). By far the greatest number of respondents said that when they wanted information about recreational drugs, they would obtain it face-to-face with the person providing the information. The second most frequently cited method was by telephone. Note that the two most popular methods are also the most confidential, as would be expected because individuals are engaging in a behavior that is not legal. The next three response categories, AIM, email, and text messaging, all had about the same number of respondents indicating that they had used this method for information.

Respondents were asked, "In the past six months, have you searched for information about a recreational drug on the Internet (e.g., marijuana, MDMA, mushrooms, cocaine, etc.)?" Of those who answered this question, 95% replied that they had sought such information. Respondents were then asked, "In general, what information about recreational drugs were you interested in getting?" Respondents were given seven possible responses to the question as to what information about recreational drugs they were interested in getting, and what follows is the percentage of respondents who answered in the affirmative to a given category: potential drug side effects (78%); potential long-term side effects (78%); how the drug makes you feel (70%); potential drug interactions (69%); how to use the drug (63%); where to get the drug (21%); and other (11%).

This is one of the most interesting set of finding in the survey. What it tells us that the Internet very much is being used for Harm Reduction, which most would agree would be a positive result of the use of the Internet. Questions about drug side effects and possible interactions with other drugs illustrate a Harm Reduction use of the Internet. On the other hand, inquiring about how a drug makes you feel and how to use the drug may be seen as information facilitating non-legal drug use. It makes sense that the Internet is not heavily being used for information as to where to obtain drugs. Since this activity is illegal, this type of information would leave those who sell drugs vulnerable.

Respondents were then asked, "Which of the following categories best describes the websites you have visited in the past six months to get information on recreational drugs?" The response categories and the percentage of respondents who sought information from websites in those categories included: drug education/harm reduction website (e.g., DanceSafe, Bluelight, Lycaeum, Erowid) (88%); sites dedicated to a specific recreational drug (49%); sites dedicated to prescription drugs (31%); health or medical website (29%); dance or club related sites (27%); drug testing kit sales site (17%); government-sponsored website (14%); drug use prevention website (7%); drug addiction treatment sites (6%); and other (4%). The results of this table are striking. First, drug education and harm reduction sites are by far the sites most visited. This finding supports the arguments of those who see the Internet as providing useful information concerning harm reduction when drugs are used. Second, prescription drug sites and health or medical websites are visited often. This either could indicate that prescription drugs are being abused or that information about drugs being taken by respondents as prescribed is being sought. Visits to government sponsored websites pale in comparison to visits to drug education/harm reduction websites, and this should be of concern to those responsible for the content of government sponsored websites. One wonders whether political constraints may be limiting science based information available on such sites.

We then inquired as to how credible the drug information found in the categories of websites is. If we look at the "very credible" response for each category, we obtain the following results: drug education/harm reduction websites (e.g., DanceSafe, Bluelight, Lycaeum, Erowid) 84%; sites dedicated to prescription drugs (48%); health or medical website (40%); sites dedicated to a specific recreational drug (38%); drug testing kit sales site (19%); drug addiction treatment sites (13%); other (9.7%); dance or club related sites (7%); and government-sponsored websites (4%).

Several findings are worthy of particular attention. First, the highest rated websites by respondents are the drug education/harm reduction websites at 84% "very credible." In fact, for this category of websites, only 1% of respondents found this category of websites as "not very credible." No other site categories came close to having the positive response rate that the drug education/harm reduction website had. At the opposite extreme, only 4% of respondents found government-sponsored websites "very credible." Almost one half of respondents (47%) of respondents found government-sponsored websites "not very credible." Respondents are indicating that they sense a bias in governmental websites, because many recreational drugs have been declared illegal. For this reason, as mentioned above, governmental websites may find it difficult to be objective concerning the potential dangers, or the lack of potential dangers, of specific drugs.

Our next question was the following: "In the future if you need information about drugs, how likely are you to go to the following sites for that information?" If we look at the "very likely" responses, we derive the following percentages for the different types of websites: drug education/harm reduction websites (e.g., DanceSafe, Bluelight, Lycaeum, Erowid) (94%); sites dedicated to a specific recreational drug (34%); sites dedicated to prescription drugs (22%);

health or medical website (20%); other (11%); (dance or club related sites (7%); drug testing kit sales site (6%); drug addiction treatment sites (3%); and government-sponsored websites (3%).

The results in Table 6 show the overwhelming preponderance of respondents saying that they will go to drug education/harm reduction websites. Two other sites, those dedicated to a specific recreational drug at 34%, sites dedicated to prescription drugs at 22%, and health or medical websites (20%) garner some support but the percentages of respondents saying that they will consult these sites in the future are much lower than the over 90% response that they are likely to consult drug education sites in the future. This indicates that almost all of the respondents indicate that they will consult drug education/harm reduction websites in the future. We conclude that first they have found such sites credible in the past, and, second, that it is to be hoped that drug education sites stay objective and scientific in the future, because the health and welfare of many people now depends on this being the case.

Our next inquiry was: "Where are you most likely to find out about a new recreational drug or drug combination?" The three most likely places to find out about a new recreational drug or drug combination were in an online chat room or discussion board (31%), on a website (30%), and at a get-together at a friend's house (22%). What this table tells us is the importance of the Internet concerning finding out about a new recreational drug or drug combination. Adding the first two percentages relating to the Internet results in 63% of respondents saying that they used the Internet to find out about a new recreational drug or drug combination. The down-side of the Internet, then for those who propose abstinence from non-legal mood enhancing substances is that knowledge about those drugs is being obtained from the Internet.

Our next inquiry was: "In the past six months, have you specifically searched the web for information on the effects and/or risks of any of the following drugs?" Responses to the question give us a feel for knowing which drugs most likely are currently being used as well as which drugs will be likely to be used in the near future. The list informs us as well as to which drugs are new enough to be not well known to the respondents, as well as those drugs not popular at the time of the survey. The list, from most searched to least searched drug is as follows: ecstasy (68%); amphetamines (56%); mushrooms (55%); LSD (51%); marijuana (45%); ketamine (44%); prescription medications (43%); herbals and supplements (41%); cocaine (40%); GBH (34%); over-the-counter drugs (30%); inhalants (28%); heroin (25%); PMA (13%); other (13%); steroids (6%); and none (5%). Ecstasy is the prime club drug, thus central to the chapters in this book, and it was the most searched drug in the survey. Note that the primary performance enhancing drugs such as steroids were the least searched.

Conclusion

The portrait that we obtain, then, from the New Drugs Internet Survey is as follows. Respondents tend to be young, male, white, well-educated, and involved in either work or school. They are likely not to be married and not to have children. Those from the United States lived throughout the United States and were not concentrated in a single region in the United States. All respondents were firmly embedded in two interconnected worlds. The first was the physical world of people, of places to meet with others face-to-face and to dance, and of a culture that includes mood enhancing substances. The second was the world of cyberspace, accessed at home, which included both information from harm reduction websites as to how to minimize the ill effects of drug use, as well as friends and acquaintances met though the Internet with whom they communicated and exchanged information. Harm reduction websites were considered more credible that government sponsored websites by a large margin. Cyberspace friends can transition into face-to-face friends as well. As mentioned before, the culture of our respondents included a combination of music, dance, and mood enhancing substances.

The survey gives us some answers to the crucial question as to whether harm reduction websites have the positive effect of reducing harm to those who use potentially dangerous drugs. The answer is clearly yes, in the sense that the use of harm reduction websites is extensive and the websites are searched for increased knowledge about drugs and how their risky side-effects can be minimized. On the other hand, though, it is also true that it is from the Internet that information about drugs, their use and their mood enhancing effects are also discovered by individuals using the Internet.

Chapter Four

Causal Factors in Drug Use: A Phenomenological Approach Based on Internet Data
Edward Murguía

This analysis takes an approach different from that ordinarily taken in studies of causes of drug use. After a review of established theories of drug use in the deviance literature, I create a composite theory of deviance based on this literature review. I then undertake a phenomenological study of causal factors in drug use, that is, I approach the subject from the point of view of what drug users themselves say about why they take drugs with data derived from a drug use harm reduction site on the Internet. Rather than seeing the different results that the two approaches yield as contradictory, I demonstrate that the two approaches produce results that fit together, resulting in a new, more complete model of causal factors of drug use.

What causes drug use? What do both causal theories of drug use and deviance theories (theories that concern themselves with deviance more broadly but include drug use) say about why individuals take drugs? Five insightful theoretical models, all with empirical support, have been developed: 1) Gottfredson and Hirschi's "self-control theory" (1990); 2) Brook's (1993) "family interactional theory"; 3) Hirschi's (1969) "social control theory"; 4) Oetting and Beauvais' (1986) "peer cluster theory"; and 5) Sykes and Matza's ([1957] 1975) "techniques of neutralization" theory.

Gottfredson and Hirschi's self-control theory is grounded on bio-psychological elements of self-control. The basis of self-control, Gottfredson and Hirschi believe, is the ability to recognize the negative consequences of deviant behavior. In their words, "the impulsive or short-sighted person fails to consider the negative or painful consequences of his acts; the insensitive person has fewer negative consequences to consider; the less intelligent person also has fewer negative consequences to consider (has less to lose)" (Gottfredson and Hirschi 1990:95).

Brook's psycho-sociological theory posits an association between low drug use and an affectionate, nonconflictual relationship with parents during childhood. She incorporates in her framework the concept of "attachment" between parent and child, defined as "the presence and quality of a warm, intimate and continuous bond between parent and child"(p.80).

According to Hirschi's social control theory, which takes a sociological approach, a delinquent is "a person relatively free of the intimate attachments, the aspirations, and the moral beliefs that bind most people to a life within the law" (p.iii). Individuals most likely to be delinquent are those not attached to normative others. Such individuals, who lack hope for achievement within conventional society and who lack conventional moral beliefs, are free from the constraints that keep most from committing deviant acts.

Oetting and Beauvais' peer cluster theory, also sociological in nature, gives central place to the influence of peers among causal factors in drug use. Peer clusters are small groups, including dyads (for example, boyfriend and girlfriend or same-sex best friends), in which drugs are made available and in which individuals learn to use them. Because deviant peer clusters prolong, sustain, and in some cases exacerbate unlawful attitudes and behavior of individuals, association with deviant peers adds an element of deviance and drug use beyond what can be explained by means of the relaxation of controls by family and school alone.

Finally, Sykes and Matza believe that techniques of neutralization contribute to delinquency. These techniques, psycho-sociological in nature, are "justifications for deviance that are seen as valid by the delinquent but not by the legal system or society at large" (p. 145). The five techniques that Sykes and Matza develop are "the denial of responsibility" (I didn't intentionally do it), "the denial of injury" (nobody really is hurt by my actions), "the denial of the victim" (the victim had it coming to him/her), "the condemnation of the condemners" (the condemners are hypocrites and are biased against me), and "the appeal to higher loyalties" (I did it for my friends).

We believe that the five causal theories for deviance presented above, rather than being competitive and contradictory, are complementary aspects of a complex phenomenon. The following scenario demonstrates how all five theoretical perspectives fit together. If the relationship between parents and bio-psychologically impulsive children is soured, as Gottfredson and Hirschi claim, and as Brook's theory postulates, there is little or no affection between these parents and their children. This leaves children free from the social bonds that limit deviance, as Hirschi's theory would have it. As these children grow into adolescence, they become oppositional youth who, as Oetting and Beauvais have shown, find others like themselves, and this coming together of oppositional youth reinforces deviant behavior among them. Within their oppositional culture, adolescents devise justifications for deviance that enable them to neutralize moral judgments coming from the larger society as Sykes and Matza believe, which leave them morally free to engage in deviant activities. To name this composite theory, I call it "control-peer-neutralization theory." Here, "control" refers both to lack of attachment to others (Hirschi 1969; Brook 1993), and to lack of self-control (Gottfredson and Hirschi 1990), "peer" refers to peer cluster theory (Oetting and Beauvais 1986) and "neutralization" refers to techniques of neutralization (Sykes and Matza [1957] 1975). I thus have a composite theory, derived from the deviance literature, as an explanation as to why individuals commit deviant acts. Let us now move to a different approach, a phenomenological approach, to examine what drug users themselves say when asked why they take illegal drugs.

Data And Methods

The data for this study are from a drug use Harm Reduction site located on the Internet. The "thread," (a series of replies and commentary to a question that begins a discussion) began with the following post (spelling and grammar are unchanged) dated July 2, 2002. Following the original post, 101 replies to this post were analyzed. The last post studied in this thread was dated October 3, 2002. Because the subject of the thread is activity that is not legal, it was decided neither to name the drug reduction site nor to give the Internet names of the originator and respondents of the thread in this study.

> *Original Post.* "Seriously, why do we do drugs? what is so great about getting high. I got this question in my head after a week of smoking weed all day and night and at the end of it all going 3rd plateau on a DXM trip. why can't we just live with the emotions that were given to us, what is so great about getting high? why do I wake up everyday and say, hey I gotta get something to f___ me up? why do I do this s____, I'm not addicted and just questioning a lifestyle I guess. please tell me why all you out there get high, and reach deep don't just tell me that you like it, I wanna know why we all choose this lifestyle. thanks.

Drug users answered the question as to why they used drugs with the following six meaningful categories of responses: 1) to satisfy an addiction; 2) for self-medication; 3) to avoid problematic reality; 4) for happiness and pleasure; 5) for friendship; 6) for insight and inspiration. The list, at least at first glance, seems to be very different from the causality posited by deviance theory. Notice that the six categories of responses have been placed in order, from drug use in the most negative context to correct painful situations, to use positively viewed by respondents for life enhancement. The texts of the responses (again, with respondent spelling and grammar unchanged) follow:

1. To Satisfy an Addiction.
Response 1.1. "I use drugs BECASUE IM ADDICTED!"

Response 1.2. "The 'don't let it control your life' is BS. . . . While not-high u think about being high so much, if not constant. So it may not control your life, but @ least it controls your toughts, well, for me it does :S."

Response 1.3. "i would have to say because im an addict, i started cause it felt good now because i hate being sober i get all depressed and s___ if i don't do anything for more then 24 hours"

Response 1.4. "I do drugs because if I don't my stomach ties up in knots, my body aches and my brain starts screaming at me. Yes I am a heroin addict. I started doing it because life is dull, full of empty promises that never deliver. Boredom sends me into major depression thinking abut how on earth could things get better when nothing interests me. I do drugs because they instantly make life nice, warm, caring and seriene. Give me a life time supply of H a comfy warm bed and I'll be quite content, sad but true."

2. For Self-Medication.
Response 2.1. "The thing with drugs is that they are, well, drugs. When we take them, we feel better, and in a way we've been trained to so that. Our concept of medicine is so reliant on drugs that we learn as children to take a pill to alleviate some kind of physical suffering. Companies market them in this fashion as well ("When the day gets rough, take . . . "). Now Tylenol is much milder than the drugs done for "recreation," but why should it be criminal to come home from a rough day at work and smoke a joint, but fine (and encouraged) to come home and pop a few Advil? In a way, we're just taking medicine when we do it, and the source of the pain differs for each person."

3. To Avoid Problematic Reality.
Response 3.1. "Hi, the reason why I do drugs is probably out of boredom and loneliness. When Im dating someone I generally never have the urge to splurge with drugs as Im getting my kicks else where. The warmth of another person liking you, and the whole relationship in all its forms is my high. When Im single, I do drugs to convince myself that Im still having some sort of fun all be it in another sense. First started using drugs experimentally, then for confidence and now just to kill time because Im bored. Nothing beats the high of being in a satisfying relationship and there is no guilt associated with it either, unlike the guilt always associated with drugs."

Response 3.2. "Me mostly boredom and recreation . . . I find when i am busy doing fun things i dont even think about drugs . . . but when im bored they are the only thing going thru my mind."

Response 3.3. "Why, cause im old,fat,ugly,intruverted,vain,self concious,got no friends (wll a couple),i have no place in life,and im bored.EXCEPT when i pop pills all of the above (and more) vanishes. And then my world then rocks."

Response 3.4. "When i first got into drugs, I was depressed and I needed an escape from reality."

4. For Friendship.
Response 4.1. "it is a great way to meet people who are not uptight. . . . , because I have met very few people who smoke and are still intolerant, anal. . . . Because my friends do it and we have a good time."

Response 4.2. "the reason i do drugs is probably because almost everyone i know does drugs and that if i didn't do them i wouldn't really have anything to do."

5. For Happiness and Pleasure.
Response 5.1. "Because my head likes it . . . and I like looking forward to something. Thats how I started smoking weed . . . I loved thinking about it all day at work . . . and well E was an every weekend thing . . . so the weekend seemed brighter . . . and fun I guess if you do it in moderation.)"

Response 5.2. "my view is that our aim of this earth is to achieve happiness and drugs are often a quick way to experiemce that for a few hours. i guess that if you talk about happiness in a scale of 1 to 10, and most people are at 8, well then by taking drugs you will just make that more erratic . . . . you will be at 28 for a few hours but then at 3 for the rest of the week . . . . thats how i kinda see it."

6. For Philosophical Understanding and Insight.
Response 6.1. "In my case is a way to open and know my mind, i do like drugs, FOR SURE yes , but i dont want em to manage or handle my life, na na na, u should know a little bit of u before take any drug, in my case I use em oftenly, sometimes for partying and sometimes to know myself, my thoughts, mind. For example: I really like Van Gogh and once I was tripping and then my whole world was like a paint of him, that day i analized that mind has amazing powerful with thoughts and things that we remember, DRUGS OPEN DOORS OF MY BRAIN, and think every time i use' em i can learn something new about my life and my environment.
The only thing I can say is not let drugs control ur life . . . try to control 'em and evrything will be diferent."

Response 6.2. "Drugs like LSD and ecstasy open new doors in your mind to experiences you can honestly not have unless you do these drugs. I really think I am a better person for experiencing hallucinogens and MDMA. I'm a writer, what can I say? There's so much inspiration behind drugs like these, why would't somebody want to do them."

Differences in the Effects of Different Drugs and Changes in Drug Use Over Time

There are two additional points that are important to make for a comprehensive study of causal factors in drug use, observations which respondents to the thread on the Internet as to causes of drug use made clear in their posts. The first is that different drugs have different effects and are used for different purposes, and the second is that for many users, drug use changes over time.

7. Differing Effects from Different Drugs

Response 7.1. "Pot: Simply to relax and think more Mushrooms: So I can explore, ponder on my life, amplify my emotions so I can further analyze them Alcohol: It runs in the family Other drugs I dont do anymore, IMHO i didnt like how E made me depressed/angry/anxious. Hopefully no more Cocaine: Had fun with it, SO done with it.

Response 7.2. "1. Marijuana enhanced music, friendship, food, laughter, rebellion 2. The odd rails of coke made me feel happy, energetic, allowed me to be Jane Q. Outlaw with my friends 3. The odd trip with mushrooms or LSD were awesome! It was like opening doors to different parts of my mind and going on adventures that were otherwise closed off. 4. I snorted meth because I loved the extra energy it gave me to do otherwise unwanted projects like housecleaning or otherwise mundane things that bored me to tears."

8. Drug Use Changes Over Time

Response 8.1. "i used drugs because i liked to experiment with new things. i continued to use drugs because they were fun, and by my use i was able to really understand and really appreciate electronic and other styles of music, which was something i lost in my mid-teen years. i stopped using drugs because i needed to have a clear head and be able to think when i needed to."

Response 8.2. "I originally started using drugs simply because I was interested. I knew that a lot of the musicians I listened to used drugs. . . . Later I got into psychedelics. I loved the way it inspired my art and music. . . . Eventually drugs started comsuming my life. I didn't want to have a job and live like everyone else. . . . . I began using everything I could get my hands on just to see what everything did. . . . After a few years, I had completely destroyed myself. . . . A few years later I was depressed and was looking for a quick cure that I could manage. So I started taking painkillers, mostly Percocet. . . . Now I'm a junky giving a half-a___ effort to get my life straight again."

These two sets of exemplars, the first demonstrating that different drugs have different effects and are used for different purposes, and the second that for many users, drug use changes over time keep the model we are developing from becoming overly simplistic. They add the elements of pointing to a variety of drugs and the necessity of knowing which drug is being used for what purpose by drug users. They also add the fact that drug users change over time relative to drugs of choice and amounts of drugs taken.

Combining the Two Approaches: Distal Discomfort Based On the Deviance Literature and Proximal Relief from Discomfort Based On a Phenomenological Approach

What is the relationship between findings from the deviance literature approach and from the phenomenological approach? Causal factors from the deviance literature are describing distal factors in drug use, pointing out the psychological/social/ physiological discomfort developed early in life as the underlying and fundamental basis of drug use. Causal factors derived from a phenomenological approach, on the other hand, emphasize proximal factors in drug use, factors more immediate than those from the drug literature that lead people to use drugs. Proximal factors derived from a phenomenological approach emphasize the relief from discomfort that drug use brings to individuals who use drugs. To clarify this statement, let us go through the factors in each model and point out why factors from the theoretical model are distal and point to causes of discomfort and why factors from a phenomenological approach are proximal and point to causes that concern themselves with relief from discomfort. Impulsivity and lack of parental attachment are the basic causes of discomfort. Lack of concern of what normative others think of their deviance frees people to perform acts of deviance. Friends also involved in deviant behavior reinforce deviant activities, and assist in devising neutralizing beliefs which allay concerns about the behavior.

From a phenomenological approach, clearly addiction, self-medication, and avoidance of problematic reality point to the use of drugs as relief from discomfort. The factor of friendship in the phenomenological model is similar to Oetting and Beauvais' peer cluster theory in the deviance literature, and emphasizes the fact that drug use often involves other people. The factors of happiness and pleasure and philosophical understanding and insight are the positive elements of drug use from the point of view of life enhancement, with the caveat that such enhancement may lead to addiction, self-medication, and avoidance of reality in the future. Finally, one element of neutralization, "the denial of injury" (nobody really is hurt by my actions), seems to be related to philosophical understanding and insight, although this is not demonstrated by the examples given here for philosophical understanding. One has a sense of this, though, in Response 2.1, a response illustrating the use of drugs for self-medication, where the respondent is making the case that there is little difference between an over-the-counter drug such as Advil and marijuana for pain relief, except that the latter has been designated an illegal drug. Finally, data from our phenomenological study make us aware that not all drugs have the same purpose; rather different drugs are used for different purposes. We are also made aware that drug use by a user is not static over time, but changes depending on the needs of the user.

The two approaches enabled us to develop a more complete model of drug use than otherwise possible. By understanding both distal and proximal factors, and with the knowledge that different drugs are used for different purposes and that drug use changes over time, we have a more complete model of causal factors of drug use than before.

References

Brook, Judith S. "Interaction Theory: Its Utility in Explaining Drug Use Behavior among African-American and Puerto Rican Youth." Pp. 79-101 in *Drug Abuse among Minority Youth: Advances in Research and Methodology,* edited by Mario R. De La Rosa and Juan-Luis Recio Adrados. NIDA Research Monograph 130. Rockville, MD: National Institute on Drug Abuse, 1993.

Gottfredson, Michael and Travis Hirschi. *A General Theory of Crime.* Stanford: Stanford University Press, 1990.

Hirschi, Travis. *Causes of Delinquency.* Berkeley: University of California Press, 1969.

Jessor, R., and S. L. Jessor. *Problem Behavior and Psychosocial Development: A Longitudinal Study of Youth.* New York: Academic Press, 1977.

Murguía, Edward, Zeng-yin Chen, and Howard B. Kaplan. "A Comparison of Causal Factors in Drug Use among Mexican Americans and Non-Hispanic Whites." *Social Science Quarterly* 79, (1998):341-60.

Oetting, E. R., and Fred Beauvais. "Peer Cluster Theory: Drugs and the Adolescent." *Journal of Counseling and Development* 65 (1986):17-22.

Sykes, Gresham M. and David Matza. "Techniques of Neutralization: A Theory of Delinquency." Pp. 141-151 in *Theories of Deviance*, edited by Stuart H. Traub and Craig B. Little. Itasca, IL: F. E. Peacock Publishers, [1957] 1975.

Chapter Five

Voluntary Use, Risk, and Online Drug Use Discourse

Melissa A. Tackett-Gibson

Many people voluntarily engage in activities that are considered risky. Ordinary "everyday risks" such as smoking and speeding are common in our daily lives. According to recent survey data (SAHMSA. 2005) recreational drug use is also common, generally voluntary and risky. Activities such as recreational drug use can have severe health consequences; and in today's world of instant online communication, individuals can easily access information about risk behaviors and health. While the study of risk and health is not new, little work has been done on the impact of internet technologies on health risk definition, assessment, and management. The extent to which the internet has become a central source of knowledge production and acquisition, it is critical to explore the impact of the media on risk construction and mitigation. This paper does so in light of recent online discourse of illicit drug use among voluntary, recreational drug users.

The Internet is a growing source of information on health in general and on drugs and drug use in particular. According to a recent study, the growth among health websites substantially outpaces that of sites dedicated to other topics (Media Metrix in Rice and Katz, 2001). Since 1998 the number of Internet users looking for information on health-related topics has more than doubled; and over 80% of all young adults ages eighteen through twenty-nine reported using the Internet to search for health and medical information (Taylor, 2002). Inasmuch as general health information is widely available online, information about

67

drugs and drug use is also easily found. A recent Internet search of active web-sites yielded over 350,000 sites related to amphetamines, 180,000 sites about Ecstasy; and over 100,000 related to crystal methamphetamine.[1] The content of such sites ranges from web "blogs" or personal diaries of drug use experiences, to government-sponsored sites aimed at reducing use and addiction. The number of sites increases daily, and with this growth individuals are now capable of dis-seminating and obtaining information quickly and anonymously. In this regard, Internet technology has changed how individuals obtain and use drug informa-tion. It has created "virtual environments" in which one may be introduced to potentially dangerous substances, obtain easy access to substances, and at the same time quickly be made aware of the harm and consequences of use.

Several prominent and reputable websites are dedicated drug use "harm re-duction." Historically following in the footsteps of grassroots health promotion efforts to reduce HIV and AIDS, online harm reduction websites emphasize the reduction of the negative consequences of drug use rather than the elimination of use. The resultant philosophy is best articulated by the International Harm Re-duction Association:

> When attempts to reduce the prevalence and frequency of risk behavior have been pursued to the limits of their apparent effectiveness, sensible public policy requires that attempts should be made to reduce the hazardousness of each re-maining risk episode.
>   (International Harm Reduction Association, 2002)

These attempts are predicated on the on the idea that drug abstinence campaigns do not reach voluntary recreational users; and therefore anti-drug campaigns do not alleviate drug use harm among a substantial number of users. While harm reduction sites often encourage sobriety, ultimately their mission revolves around the dissemination of "unbiased" knowledge from which drug users and potential users can base rational use decisions. With increased understanding of drugs and their harmful effects the risks associated with use can be diminished. As one harm reduction advocate states:

> Approaches such as [those proposed by Harm Reduction] are radical in that they require the acknowledgement and acceptance that some people are not ready to give up high risk behavior. Making connections by helping them in other ways can reduce harm and open the door to further intervention.
>   (Westermeyer, n.d.)

This paper considers drug use risk discourse in light of key concepts from Beck's *Risk Society* (1992). First it examines how individuals involved in rela-tively high risk behaviors, such as illicit drug use, confront health risk and spe-cifically how online discussion helps define and mediate use risk and conse-quences. In other words, the current study examines what constitutes risk in a "virtual" setting and the means through which discussion participants attempt to reconcile knowledge of harm and drug use risks with their voluntary use. Thus it examines the socio-cultural development and mediation of risk knowledge online. Secondly it explores the contest—that has uniquely grown out of online communication—between "lay" assessments of risk and "expert" accounts of drug use harm. It addresses the ways in online discussion participants offer risk definitions that compete with hegemonic risk discourse, the impact of "lay" accounts on prevailing and "authoritative" perceptions of risk, and ultimately, the impact on drug use prevention policy.

The Internet, Drugs, and Risk Society

Theories of risk in modern society have been widely applied to issues of health and well-being. For example, Castaneda (2000) examined the health rumors and media of risk reproduction (2000); Beck-Gernsheim (2000) looked at cultural perceptions of health and responsibility to public acceptance of genome therapies. Others still have examined the risk and harm in the rhetoric of young mothers, and assessed drug production and the practice of medicine in light of Risk Society (Murphy, 2000; and Moldrup and Morgall, 2001 respectively). Similarly, social scientists have examined voluntary risk taking and high risk behaviors (Lupton and Tulloch, 2002; Stranger, 2001; Murphy, 2000; Natalier, 2001).

Recent theories of risk are particularly relevant to the study of online harm reduction discourse. In particular, sociologist Ulrich Beck, in his theoretical discussions of *risk society* (Beck, 1992), considers risk the product of uncertainties of life in late modernity, that risks follow the attempts to reconcile uncertainty. Risks are not only perceived and confronted, calculated against gains and benefits; but they are more often unseen, unnoticed, and illusive. They are the unintentional consequences of modernity itself (Beck, 1992; Beck, 1994). As Castaneda explains, risks—the result of technological change—infuse all aspects of society. Risks are imbued with meaning to the extent that they are unable to be articulated and critically assessed by experts. Perceived consequences increase to the extent that uncertainty is prevalent, and intensify to the extent that risk crosses global boundaries and move beyond expected predictability (2000).

Knowledge, its absence and development, plays an important role in *risk society* (Adam and Van Loon, 2000; Beck, 1994; Giddens, 1994). Beck contends that modern society encounters the risks of technological advances through the production of knowledge. Institutions such as the media and legal and scientific professions have a substantial role in the production of risk definition and management. Risk knowledge can be "changed, magnified, dramatized, or minimized within knowledge, and to that extent [risks are] particularly open to social definition and construction" (Beck, 1992: 23). In this sense risk definition, harm assessment, and methods of risk management are frequently contested. Various socio-political interests collide as they seek to manage risk, modify behavior, and allocate consequences. For example, if lung cancer risks are to be managed and reduced, behaviors that contribute to that risk such as smoking, should be modified; if an outbreak of SARS confronts a location, efforts to isolate that disease such as reducing the number of flights to that location, are employed. It is in this way that risk definition and management is often a mechanism of social control, one in which the authority to produce credible knowledge and redefine risk is challenged.

Beck would consider the "contestation of risk" an attribute of modern reflexive societies in which not only industrial society must confront the technological products of its own creation, but also where individual lives become lived choices and self-regulation (Beck-Gernsheim, 1996:150). Following Beck and Giddens, Adkins explains that in *risk society* uncertainty and risk are so articulated that people are asked to reflect upon their ability to control risk. They are expected to be individual managers of personal risk, to be responsible for risk consequences, often without the necessary resources necessary to do so. In this context:

> a planning of one's own biography and social relations where the opening up and dis-integration of industrial society and the untying of individuals from its rules, norms and expectations such as class and status compel people not only to create and invent their own certainties, forms of authority and regulation but also to create and invent their own self-identities and themselves as individuals. In risk society the standard biography therefore becomes a chosen, do-it-yourself, reflexive biography mediated by categories of risk (Adkins, 2001).

Not only are individuals required to manage lives full of risk, but they must also assess extensive amounts of knowledge related to risk and consequence in order to do so. Experts grow up to assist in this effort (Giddens, 1994).

Theoretically harm reduction websites can be viewed as an artifact of Beck's "risk society." As Adam and Van Loon note, the process of risk definition is highly political (2000). Therefore, various and contested assessments of risk and risk consequence struggle for public legitimacy. With the development of the internet, and more importantly the growth of online communities, various groups previously excluded from this contest, can readily participate. They can contribute competing views of risk and harm counter to prevailing perceptions. They can establish easily accessible "associations" of experts, and establish methods of managing the consequences of risky behavior.

Ironically it is the ease of use and proliferation of online information that can be perceived as the risk of the Internet. The risk of Internet use is uncertain and unpredictable. The uses of its products—information—are unknown, as are often the sources of that information. Political efforts to control its content and minimize the risks of "misinformation" are ineffectual due to its global reach. Readily available information, which may or may not be "real" or "true," presents risk misinformation and risk introduction; both of which may contribute to the initiation to or continuation of risky behavior. For example, harm reduction information online may deter drug abuse and ultimately contribute to a decreased incidence of injury and death due to abuse. However, the ready availability of drug use information might also entice individuals to use harmful substances, and inadvertently intensify the harmful effects of drug use. People are not only able to explore risks but also those behaviors, positive and negative, associated with risk.

Similarly, prescription drugs and illicit substances can be viewed as products of *risk society*. As Moldrup and Morgall, argue all modern drugs are high-tech artifacts with high-tech risks (2001). As they trace the production, marketing, and use of fluoxetine (Prozac), they conclude that while marketing drives global demand for a drug it also delays the recognition of its side-effect profile. The actual risks associated with a substance may only appear many years after the drug has gained wide, even global, appeal and use. Personal experience with drug risk is, as Beck describes, "second-hand non-experience"—often experienced only through media accounts, interpretations of scientific data produced by the manufacturers, or socially mediated information from experts (2001:69). Ultimately, even those systemic mechanisms established to prevent risk and harm, such as the FDA, can do little to protect the public from unseen drug risks. In short, "the consequences of drugs are invisible to individuals . . . therefore [they are] highly dependent on cultural and scientific interpretation" (2001:71). Accordingly, drug use risks, like others in *risk society*, create a unique space for "lay" and "expert" opinion and interpretation of consequence. It is this space that is readily found online in harm reduction discourse.

The Research Setting and Online Communities

Research data were obtained from transcripts of online threaded discussions that took place at a large popular drug use information website, for the purposes of this paper it will be referred to as "DrugSite." The site, founded as a means of providing drug information so as to reduce dangerous drug use practices, offers archives and discussion forums related to specific substances, methods of use, and dangerous drug interactions. The site was founded on the principles of harm reduction. As stated in its online introduction to new members:

> If someone is educated in the general principles of harm reduction, they increase the odds that their drug use will not lead to short-term disasters or long—term negative consequences. . . . [W]hile [DrugSite] advocates harm reduction and attempts to eliminate misinformation, there is definitely no such thing as safe drug use. [DrugSite] does not condone or condemn drug use. We believe that accurate information and encouraging personal responsibility are more helpful than using scare tactics or distorting the truth. [We give] you the opportunity to learn from your peers about drugs in an informal environment, but ultimately it is up to you to make your own decisions about drug use.
> (DrugSite "Our Mission," 2003)

It was described by its administrators as one of the largest online sources of drug and drug safety information. It was often offered as a link from similar drug-related websites as a credible source of drug information. At the start of research, DrugSite had over 35,000 registered members, 75,000 active threads of discussions, and more than 850,000 individual posts. Overall it obtained a large amount of daily traffic as registered members and unregistered guests (e.g., "lurkers") participated in more than thirty distinct discussion forums. Discussion forums at the site varied from those dedicated to drug information and harm reduction to those related to current events, the arts, and local interests. According to site statistics, activity in drug-related forums accounted for only 18% of the site's daily traffic. Site traffic was predominantly concentrated in forums dealing with miscellaneous chat, politics, music events, and the arts. Ten of the site's forums were specifically dedicated to health, drug use, and drug use safety at the time of the study.

Three forums, "Pill Primer," "Reflections," and "The Harder Stuff," were selected for the focus of this study, each primarily dealt with issues of drug use and drug effects. Based on site use patterns these forums were among the most active of the ten drug related forums. However, they accounted for only approximately 10% of the traffic on the entire site. In general, most of those participating in the forums selected for this study directly acknowledged some level of drug use. Most referred to substance use on weekends or only at parties; a few acknowledged heavy daily use or self-described addiction. They were older adolescents and young adults who were to a large extent actively engaged in educational and occupational activities.

The first of the forums examined in this paper, "Pill Primer," dealt mainly with topics of interest to new substance users—typically users of ecstasy and other so called "club drugs." The majority of the participants in that forum were newer members of the website and they discussed drug interactions, adequate dosages for "good rolls," "good" drug combinations, and problems/unanticipated effects of drug use. Similarly, participants of "The Harder Stuff" discussed methods of drug use, interactions and effects; however, most on this board described experiences with non-medical use of prescription drugs (e.g., opiates, benzodiazepines, etc.), heroin, various stimulants, and other drugs. Active participants on this forum were in large part very well educated about the drugs that they used and were long-term members of the website. They were also often described as "hard core users." Participants in the "Reflections" forum were also typically long-term members of the site. They focused discussion on personal and emotional problems attributed to their drug use. Members of this forum often posted poems, diary entries, and other personal narratives dealing with irresponsible use, introspection about use and abuse, and recovery from addiction.

Much has been written about the attributes and interactions of Internet communities (Rheingold, 1993; Bell, 2001; Wellman and Gulia, 1999; Haythornthwaite, 2001; Ward, 1999). Overall, a defining characteristic of DrugSite, like many other "virtual communities," was its emphasis on sharing experiences and supporting the well-being of its members (Rheingold, 1993). Relationships at DrugSite, like those in face-to-face communities were often supportive, reciprocal and at times contentious (Wellman and Gulia, 1999). Members used their virtual encounters as a springboard to other contacts, both offsite (via Internet chat relay, email, or telephone) and face-to-face. "Meet-ups" were organized through the site and references to face-to-face encounters were frequently seen on discussion boards. Members of DrugSite were proud of both the online and offline community that had become established. As one site administrator explained,

> [P]eople make relationships from the boards. It's the medium of our generation. There is such a closeness in the community there. The founders [of the site] didn't just create a harm reduction site, they created a community. I have maintained relationships with people all over the board with email, [instant messaging and the] message boards. Most of the people on my [cell] phone [directory] are DrugSiters that I know on a first name basis—I call them all the time . . . . DrugSite is most interesting not so much because of the drug conversations but because of the community presence online.
> (DanZerD, DrugSite Administrator, 2002)

Inasmuch as there was a site-wide community of "DrugSiters," over the course of the research it also became clear that specific individuals frequented specific forums. While there was crossover between forums (usually by the most active site members and forum moderators), DrugSite members consistently participated in only a few of site's forum discussions. In fact in many ways, each forum constituted its own virtual community with moderator/experts and "committed" participants. Specific forums had reputations as being "too introspective," a "bunch o' newbies," or "over the edge." On occasion forum participants would remark on their unique communal identity within the larger site. For example, members of The Harder Stuff were identified uniquely as HSrs; similarly they organized annual face-to-face meetings, to which members of other forums were not invited.

The unique membership and purpose of each forum was also recognized and generally respected by site members as well. In defending his contribution to a forum discussion, one moderator commented that:

> [B]ack when I would make posts to the [Ecstacy] forums I used to think about [harm reduction and new drug users] in my response, now [at The Harder Stuff] I just share various experiences in my life. . .You will notice my lack of posts in Health and Wellness—I honestly feel like yelling at half the posters and telling them to quit their bitching and just get fucked up. . . but thats not in the spirit of [that] forum
> (NoesticO, July 2002)

Research Data and Methods

In July 2002, I contacted the website's administrators to inform them of our research agenda and to request their consent to collect data from the site. While they agreed to assist in the study, they requested that my participation in the forums be limited to that of an "anonymous lurker." Therefore, I followed active discourse, read and archived transcripts of threaded discussions, but did not post comments or questions to the board or initiate offsite contact with forum members. Several discussions took place between October 2001 and March 2003 that dealt with issues related to drug use risks, potential harm, and negative consequences of use. Six representative discussions were selected for analysis. Both males and females participated in the selected discussions, although due to the anonymous nature of online communication it was difficult to determine the number of men or women that contributed to the discourse. While in general those posting on the "Pill Primer" forum were new members or unregistered guests at the site, each specific discussion included input from site visitors, "newbies," regular participants, moderators, and administrators. In all 158 individuals contributed to the selected discussions. Again, all of those that posted directly acknowledged recent personal drug use or described the use of a close friend. In the course of discussion they indicated a wide variety of drug use experiences (i.e., length of time drugs used, frequency of use, types of drugs used).

My analyses of the transcripts centered on risk related discourse—its narrative and characterizations of drug use risks, the assessments of that risk, and methods of risk mitigation. The following discussion details that analysis.

Drug Use as Pleasure and Harm

Lupton and Tulloch found in their work that risk-taking, while often associated with harm and danger, is often regarded as pleasurable, and/or an important means of self-actualization (Lupton and Tulloch, 2002). Similar views of risk-taking behaviors are also found in DrugSite discourse. To many forum participants the resultant effects of drug use were viewed as desirable, enjoyable, and with some users, necessary. For example, in two discussions of how drug use has changed their lives, several site members considered the positive consequences of their behavior. They attribute greater social skills, increased creativity, a broader life outlook, and more appreciation for their individuality. For these members drug use was pleasurable, rewarding, and beneficial.

> i gain a lot from the experience. words seem to work when im peaking, i write things i would have never been able to, music. I've been doing dxm now for, mmm, like 4 months about. every 5th day i've been taking it, or about 4-5 times a month. i really enjoy its effects, everything feels right and in place.
>   Adriantics (September, 2002)

> i used to have horrible problems socializing and just trouble getting along with

others in general,until i began expirimenting with mdma. when on it i could express my self in ways i couldnt before and when i expressed these things while high i got expierince of how to express my self when not.

Butterfly_X (March, 2003)

I have done mostly psychedelics and MDXX, and I feel that these drugs allow you to almost drop yourself into a different mode of existence (sic), apart from the one that we have been brought up in, ie society. By removing yourself from the stream of people doing what they normaly do in their daily lives, you can look at just what it is they do from a completely different perspective. . . . Before taking these drugs, I had no idea that there was more to "it." I had been brought up just like everyone else to believe that what is is, and you should just accept it. But that is a very wrong assumption based on limited facts and observations.

The Shaman (March, 2003)

After use I have more confidence about myself, I feel I know what I need to do in the world. I also feel as though I either have more control of my life, or less (but accept it) . . . hard to explain. Drug use has helped me get a handle of some things, and helped me through some difficult situations. Where as some may see a psychiatrist, I trip, and have that 1 on 1 convo with myself.

MadDog_3093 (March, 2003)

In addition to discourse that emphasized the benefits and pleasures of drug use, there are many instances where those benefits are weighed against the negative outcomes of use. The enjoyment of use is contrasted with its consequences as site members remind each other of the problems of negative health effects and irresponsible use. They indicate that while drug use is enjoyable, it is also potentially dangerous. As one participant reminded Adriantics, "dxm is probably not a good drug to do continuously . . . just like any other drug . . . your brain will thank you when you're 50 and still able to go to the bathroom by choice" (CaverRaver, September 2002). Drug use was characterized as both pleasurable and potentially harmful. As other participants noted:

[Drugs] A tool;yes. Fucking good hedonistic fun; yes ☺ As well it does appeal to the "lets see where this little fella takes us tonite" part of my brain. All in all e has taught me to enjoy this shit we call life when I am sober . . . to be able to do the things when scattered are easy, but sober they are hard. But the more I do it, the more I realize (sic) how negative it can be. a catch 22 perhaps? much like having a doomsday machine and not telling anyone about it.

#egger#Sam (August, 2002)

I'm no angel, I drink, smoke weed, to E, do coke, but I don't delude myself into thinking these are COMPLETELY SAFE behaviors. I just know were my level of acceptable risk is, and try to live within it.

Citizen_K (March, 2003)

Managing Drug Use Risk

Since drug use was potentially harmful, its risks could be balanced against its negative health and social effects. For some the risks of use were worth the pleasurable experiences they provided—the positive outcomes of use outweighed the potential harm. To these DrugSiters the consequences of use were not immediate. The harm that might come from drugs or their incorrect use was projected to the future—that the effects of use would be experienced in old age, and that they might be acceptable. One forum participant concluded that he would continue to take drugs even if he knew that the use would lead to long-term negative health effects. He described his decision to use drugs as an immediate one; in contrast he indicated that he would consider drug use risks in the future.

> i don't plan my life a week in advance, let alone make decisions which will affect my life 40 years in advance. Therefore I'd probably still take drugs knowing full well they'd f——me up in the long term. . . . however, the answer I've given is really a selfish one. I plan to be a grandfather one day and would I want to be a vegetable at 65 and not be able to spend time with my family and friends? Probably not. . . . My final answer is: I'll think about it next week.
> Jetliner (October 2002)

Another explained:

> Im largely for "live for the now" so [the risks] aren't really stopping me . . . when I eat a hamburger I think of how it will affect me in the now, not in 40 years down with bad health, and it's the same with drugs.
> TaTiana (October, 2002)

Similarly one DrugSiter described what he would do if he found "clear evidence" that MDMA caused brain damage. He outlines a process of tallying risk and harm against drug use benefits. To this participant, enjoying drug use was worth a certain amount of harm that might come from that use.

> [If I knew for sure] I'd weigh up all the benefits, compare then to the new risks I just discovered, and decide whether it was worth it. For most drugs I don't think the risk would be worth it because the benefits I get aren't that huge, but with MDMA I think the risks would have to be fairly significant, because the benefits I received and could possibly receive again are substantially higher.
> NeoPlastic (October 2002)

To Neoplastic drug use was risky and had harmful effects. However, the negative effects of use could be diminished by gaining drug use knowledge and rationally weighing use risks against benefits. Similarly, other DrugSiters argued that risks should be immediately confronted and managed. By learning about drug use risk, participants could use "smartly" and potentially mitigate harmful consequences. A clear understanding of substances and their effects could reduce their negative impacts. In each of the forums reviewed, site members expressed the need to evaluate risk and reduce it through knowledgeable and responsible drug use. Site members, particularly older DrugSiters, frequently instructed "newbies" on drug use methods and safe practices. Discussions often included warnings about the use of particular drugs and drug combinations. New site members were commonly referred to other websites or DrugSite FAQ documents that provided detailed information about drug effects, safe methods of use, and harmful combinations. For example, during discussions of the use of PCP with Ecstacy, forum participants provided narratives of enjoyable experiences with the drugs. They also offered the following warnings:

> PCP is very strong and should not be played with, please use caution when handling and taking it
> > MysteriousD (September 2003)

> MDMA can give you scary time if you don't know how to do it or to keep it under control. otherwhise is fine.
> > ToxicMutherF———(June 2003)

NoesticO, a DrugSite member well known for his extensive knowledge of drugs—and his repeated use of hard substances—cautioned newer users to learn more about drugs before they began use. As he instructed another forum participant about appropriate methods and dosages for morphine use he warned:

> [the dosages given] are all VERY CONSERVATIVE and may be seen as LOW or not enough to "get you mad fucked up"—but you must start small and work your way up, as you don't want yer first experience with morphine to be your last . . . Treat her with respect and she will make you very happy!
> > NoesticO (March 2003)

Similarly, during a discussion of a friend's overdose, another well regarded site member reminded users to learn about the drugs they use and use responsibly:

> I don't want to sound like a preacher, but in my experience drugs in all forms (legal/illegal) need to be respected and understood. . . . Read, learn and understand anything you put in your body. If your dumb enough to inject heroin in your veins, please realize the stupidity and danger you put yourself in. If you think "E" is no big deal, step back and read. . . . [F]ully understand the risks. . . . [Drugs] are an experience in a life with SMART, LOGICAL, decisions.
> > Zeus (February 2002)

Contested Knowledge of Risk and Harm

Gaining drug use knowledge (in order to reduce harmful consequences) wasimportant to forum participants. Inasmuch as it was emphasized as the website's primary mission, the value of knowledge was confirmed in forum discourse. As stated earlier, however, definitions of risks and harm are a product of social processes. Risk knowledge is constructed, ever-changing, with various experts providing assessments of "real" risks and risk mitigation. In this sense, definitions of risky behaviors, their consequences, and the means of risk reduction are contested with various "authoritative knowledges" of risk competing for legitimacy.

This contest, between "lay" and "expert" risk definition, is clearly played out in forum discourse. While risk knowledge is critical to drug use harm reduction, only a few sources of that knowledge are considered legitimate. Participants often referred to drug education programs as "propaganda" that offered distorted facts about the harm associated with drug use. The media were accused of misreporting scientific studies, or capitalizing on drug use associated deaths to exaggerate the consequences of use. For instance, one site member describes his misconceptions of drugs in the past. In doing so he faults the media and programs, such as D.A.R.E., for the misinformation.

> I used to think that marijuana made you significantly stupider . . . blame that one on Hollywood and their portrayal of stoners.

> I once thought that acid was supposed to be what anticholinergics and stuff like jimson weed ARE. Complete disconnection from reality, delusions, illusions of grandeur, etc. This goes for PCP, as well . . . blame it on DARE.

> I was taught, that marijuana is the gateway drug and the second you start smoking it, you'll begin thinking about doing harder drugs like heroin and cocaine . . . DARE, again!

> I thought only trashy, grungy people did drugs . . . videos shown by various sources of propaganda . . . INCLUDING DARE.

> I thought MDMA was the date rape drug . . . oh, look, DARE . . .
> WhiteboyFunk (March 2003)

Similarly, others criticized D.A.R.E. and other school-based education programs for offering misinformation, and even lying about drug use and harmful consequences. To these participants coordinated drug use reduction efforts had little credibility and were poor sources of drug information:

> when DARE rolled around in 6th grade I thought that people who did drugs were all gang members who had nothing better to do all day long than find ways to force other people to take drugs, sometimes at gunpoint. I was always worried that someone would put a gun to my head and then I'd have to choose between my life and taking drugs. . . . Oh yes . . . and in DARE they told us that people made acid in their toilets and sometimes forgot to flush them. No joke.
> Paperdoll (March 2003)

we did get the teacher when she said its commen for people to OD on cannabis.
"No it isn't, its never happened."
"Yes it happens all the time"
friend shuffles about in bag, pulls out article from newspaper about Cannabis
"See, this says noone has ever OD'd on it"
teacher losses all credibility with the room

Cruel I know, but its the only way they'll stop telling outright lies.
AlphaDog (March 2003)

In contrast, current scientific and medical studies, website archives, and other drug information websites were frequently cited as reliable sources of information regarding drug use risks. In fact most of the references to "accurate," "unbiased" information about harm pointed to an online source. In response to a question about morphine dosages, one forum moderator suggested that he, "do a search for Morphine at erowid.com." Another pointed a new user to "our FAQ pages." Drug information from RxList.com was frequently used in drug discussions, and similarly, participants referenced clinical study data from Medline's archive. Entitled "Myth" a forum member posted a link to the web archives of the Multidisciplinary Association for Psychedelic Studies (MAPS):

> Ecstasy burns holes in your brain.
> Status: Untrue
> Source: Oprah; MTV's Special on Ecstasy, 11/28/2000.
> Rebuttal: Rick Doblin, PhD—see http://www.maps.org/media/mtvclarify.html
> AlwaYs at Ease (June 2003)

Online discourse related to personal drug experiences was also frequently considered a credible source of information about drug use and consequences. In fact other trusted drug users were often viewed as the most reliable sources of "unbiased" knowledge. Their experiences were credible due to their nonjudgmental stance toward drug use and their willingness to gain personal experience with a variety of substances. As one board moderator describes:

> When we suggested that is was a very bad idea to abuse Halothane the posters believed us. After all the same people who have no problem shooting street heroin are telling them not to mess with the stuff . . . it must be fucking dangerous!
> BuSh_N (July 2002)

In response to a DrugSiter's description of a friend's overdose from a new illegal substance, another board member described her own experience with the drug. In doing so she acknowledged the impact of the story on her drug use decisions:

> Sorry to hear [about the overdose]. Those [pills] have made there way up to Philly. There all over now. I myself and a lot of friends were going to try them .
> . . I was waiting to do this. After reading this thread I'm now having second

thoughts.
                    PanamaJill (July 2002)

When asked what it would actually take for him to quit or moderate his drug use, one participant acknowledged that information that indicates that drug use may do some physical and psychological harm is important but not convincing. For his use to be dramatically impacted he ultimately needed confirmation of harm from personal experience:

> I'd need to see something hard like someone who is fucked up from drug taking, and the like, from when they were younger. Itd probably take more than that as well, although it would certainly make me think.
>   DogDay (September 2002)

Conclusion

Online discourse related to drug use and consequence highlights several important attributes of risk assessment within a harm reduction community. Primarily, the discourse demonstrates the self-reflexive nature of voluntary risk-taking. In this online forum, risk-taking behaviors are assessed, analyzed and described in terms of benefits and harm. To site participants, personal use is not only pleasurable, but it is also potentially dangerous. This pleasure can be weighed and measure against potential harm. Risks in this context are calculable, measurable and quantifiable. Accordingly it is one's personal responsibility to assess and manage harm and pleasure.

Not only does the discourse indicate a high degree of consideration of behavior and consequence, but it illustrates the importance of knowledge and the development of expert individuals and "sites." Risks are only manageable with adequate information and knowledge. Such knowledge is contested, to the extent that the identification of "truth" is problematic and political. In this case, participants value the knowledge and experience of other site members. They provide details of behavior and its resultant consequences so that others can use that information to weigh use and risks. They also value other online avenues of information, such as drug information archives, drug research sites, and other discussion boards. The community readily rejects the accuracy of other publicly accessible sources of knowledge such as school education programs and government efforts to reduce drug use.

Drug use discourse demonstrates several theoretical characteristics of risk society—primarily that high-tech methods are used to assess the risks of high-tech artifacts. That knowledge is mediated through social processes that are in the instance a product of online "virtual" interaction. In a harm reduction community risks are "second-hand non-experienced." They are defined through a process of interaction reliant on interpretation of scientific data and personal accounts of use. Risks are unseen, left to be experienced at a future time—when it is too late to mediate risk consequence. They are illusive and undefined. As one participant indicated "we really won't know if we're causing brain damage . . . we're the guinea pigs in this drug experiment."

Notes

1. Active websites were defined as those that had been updated in the past three months.

References

Adam, B., U. Beck, and J. Van Loon. *The Risk Society and Beyond: Critical Issues for Social Theory.* Sage Publications: Thousand Oaks, CA. (eds.) 2000.

Adkins, L. "Risk Culture, Self-Reflexivity and the Making of Sexual Hierarchies." *Body and Science* Vol. 7(1) 2001: 35-55.

Beck, U. *Risk Society: Towards a New Modernity.* Sage Publications: Thousand Oaks, CA, 1992.

———. "The Reinvention of Politics: Towards a Theory of Reflexive Modernization." in *Reflexive Modernization: Politics, Tradition, and Aesthetics in the Social Order,* edited by U. Beck, A. Giddens, and S. Lash, Stanford University Press: Stanford, CA, 1994.

Beck-Gernsheim, E. "Health and Responsibility: From Social Change to Technological Change and Vice Versa." in *The Risk Society and Beyond: Critical Issues for Social Theory,* edited by B. Adam, U. Beck, and J. Van Loon. Sage Publications: Thousand Oaks, CA, 2000.

Bell, D. "An Introduction to Cybercultures." Routledge: New York. *Body and Society,* 7 no. 1(2001):35-55.

Castaneda, C. "Child Organ Stealing Stories: Risk, Rumor, and Reproductive Technologies." in *The Risk Society and Beyond: Critical Issues for Social Theory,* edited by B. Adam, U. Beck, and J. Van Loon. Sage Publications: Thousand Oaks, CA, 2000.

Giddens, A. "Living in a Post-Traditional Society." in *Reflexive Modernization: Politics, Tradition, and Aesthetics in the Social Order,* edited by U. Beck, A. Giddens, and S. Lash. Stanford University Press: Stanford, CA, 1994.

Haythornthwaite, C. "Introduction: The internet in everyday life." *American Behavioral Scientist,* 45, no 3 (2001):363-382.

International Harm Reduction Association. "What is Harm Reduction?" 2002. <http://www.ihra.org> (5 May 2003).

Lupton, D. and J. Tulloch. "'Life Would be Pretty Dull Without Risk': Voluntary Risk-Taking and its Pleasures." *Health, Risk and Society,* 4, no. 2 (2002): 113-24.

Lupton, D. and J. Tulloch. "'Risk is a Part of Your Life': Risk Epistemologies Among a Group of Australians." *Sociology,* 36, no.2 (2002):317-34.

Moldrup, C. and Morgall, J. "Risk Society—Reconsidered in a Drug Context." *Health, Risk, and Society.* 3, no. 1 (2001):59-74.

Murphy, E. Risk, "Responsibility, and Rhetoric in Infant Feeding." *Journal of Contemporary Ethnography.* 29, no. 3 (2000):291-325.

Natalier, K. 2001. "Motorcyclists' interpretations of risk and hazard." *Journal of Sociology*, 37, no.1 (2001):65-80.

Rheingold, H. *Virtual Community: Homesteading on the Electronic Frontier.* Addison Wesley: Reading, MA, 1993.

Rice, R. and J. Katz. *The Internet and Health Communication: Experiences and Expectations.* Sage Publications: Thousand Oaks, CA, 2001.

Substance Abuse and Mental Health Services Administration. "Overview of Findings from the 2004 National Survey on Drug Use and Health." Office of Applied Studies, NSDUH Series H-27, DHHS Publication No. SMA 05-4061. Rockville, MD. 2005.

Smith, M. and P. Kollack. *Communities in Cyberspace.* Routledge: New York, 1999.

Stranger, M. "The Aesthetics of Risk: A Study of Surfing." *International Review of the Study of Sport*, 34, no. 3 (1999.): 265-76.

Taylor, H. "Internet Penetration at 66% of Adults (137 Million) Nationwide." *Harris Poll #18, April 17, 2002.* 2002. <http://www.harrisinteractive.com/harris_poll/index.asp ?PID=299> (15 Jan. 2003).

Ward, K. Cyber-ethnography and the emergence of the virtually new community. *Journal of Information Technology*, 14 (1999):95-105.

Wellman, B. and Gulia, M. "Virtual Communities as Communities: Net Surfers Don't Ride Alone." in *Communities in Cyberspace*, edited by M. Smith and P. Kollack. Routledge: New York, 1999.

Westermeyer, R.W. n.d. "Reducing Harm: A Very Good Idea." <www.habitsmart.com> (6 May 2003).

Chapter Six

Deterrence of Harm to Self: A Study of Online Rhetoric
Azzurra Crispino

United States drug policy is driven primarily by a combination of two theories of drug use, rational choice theory and deterrence theory, which will be described below.[1] In turn, these theories are founded on the premise that "all individuals choose to obey or violate the law by a rational calculation of the risk of pain versus potential pleasure derived from the act." (Akers 2000, p.16) The received view is that if the government is to deter drug use, the best way to do it is to enforce formal and informal sanctions exactly stringent enough to overcome the apparent gain that one would experience from using a drug. In describing the rationale behind drug policy, prominent drug theorist Erich Goode states "most observers and combatants [of the drug war] are not consequentialists" but rather moralists (Goode 1999, p. 96). That is to say, the people who are making drug policy are not primarily concerned with the consequences of drug use but rather with the morality of using drugs. The argument of the moralists is that drug use is morally wrong, and hence it is the responsibility of government to dissuade people from using drugs, by whatever means necessary. But what of the users? Goode makes no mention of how people make the decision of whether to use drugs. The scope of this article is to articulate how real people make the decision of whether or not to use drugs, and see if there are any differences between how *real people* use *real drugs*, as opposed to how *policy makers* envision people using drugs. Whatever discrepancies found should be incorporated into drug policy, if we expect drug policy to be effective.

In this paper, I will argue that formal and informal sanctions do not play as significant a role in shaping the decision making process with regard to drug use as traditionally thought. By analyzing the discourse both of parents attempting to deter their children from using drugs, and drug users themselves as garnered from two websites, we will see that both sides of the drug use controversy are far more concerned with what they see as the intrinsic effects of the drug use, or what I will call causal-empirical deterrence, than with sanctions. Causal-empirical deterrence is defined broadly as the negative effects one experiences from an action intrinsic to the action itself, and not due to an external agent. I will conclude by arguing that causal-empirical deterrence, which has some overlap with formal sanctions but is not entirely encompassed by the concept, is actually more often used in the discourse of those making the decision of whether to use drugs than sanctions. This is not to say rational choice theory and deterrence theory do not capture some of how people make decisions about drug use, but it is to say that it is not all of it. Finally, based on my analysis, I offer policy recommendations concerning what I believe is the optimal approach to discourage use of those illegal drugs that are truly physiologically and psychologically harmful.

Deterrence Theory, Rational Choice Theory, and Drug Use

Rational choice theory and deterrence theory are two separate theories, but they are often used in conjunction when creating criminal policy, as is the case in drug laws. As Ronald Akers explains in his book *Criminological Theories*, Rational Choice Theory is based on the view of human nature that "people will make rational decisions based on the extent to which they expect the choice to maximize their profits or benefits and minimize their costs or losses." (Akers 2000, p.23) Therefore, if the government wants to deter its citizens from choosing a certain action (such as using an illicit substance) the government's role is to maximize the cost or loss incurred by the action and minimize its benefit. Deterrence theory outlines two types of deterrence, specific and general, where specific deterrence targets those who are caught and punished for a crime, and general deterrence refers to the rest of the population. General deterrence requires that people are aware of and understand the laws. Furthermore, both types of deterrence are based on maximizing the severity, certainty and celerity of punishment. Concerning severity of punishment, the punishment for a crime must be severe enough to overcome the benefits gained by the crime in order to deter, but will be unjust if it is overly severe. Certainty of punishment is the probability of being apprehended and punished for a crime. Celerity of punishment refers to the speed with which one is apprehended and prosecuted.

Although all three are required in order for deterrence to work, the severity and certainty of punishment required are interlinked: "the more severe the punishment, the less likely it is to be applied; and the less certain the punishment, the more severe it must be to deter crime." (Akers 2000, p.17) Celerity of punishment is not usually considered, though one way it is incorporated into the legal system is through the right to a speedy trial guaranteed in the Constitution.

The severity of punishment requirement of deterrence theory was the sole theoretical justification for the Reagan Administration's decision to push for mandatory sentencing of drug crimes. The justification was that although the certainty of the apprehension for drug crimes could not be guaranteed, the severity of the punishment would be enough of a deterrent. This had not previously been the case when judges were allowed to dictate the punishment in an *ad hoc* manner. The increased push for higher arrest rates of drug users is also a direct extension of deterrence theory.

Education is also a major component of current drug policy, although educational anti-drug programs are also based in deterrence principles, as educational programs focus on the negative effects of drug use as a deterring effect, along with targeting adolescents with education and advertising that highlights consequences of drug convictions. The inculcational educational aspect is a key component of deterrence and rational choice theories, although it is not explicitly a part of them. The threat of legal punishment, even if it does not reflect the actual law, should have a deterrent effect.

Recent studies seem to indicate, however, that deterrence has more to do with informal sanctions than with formal sanctions. Formal sanctions refer to the government's direct way of maximizing the cost for an action, such as prison terms. Informal sanctions are on a more community level, and can include parental sanctions (getting grounded, for instance) or exclusion from social groups. For example, juveniles are more concerned with their parents' reaction to both to the crime itself and to the sanctions that their children will receive than with the sanctions themselves: "the informal sanctions have an independent effect on delinquent behavior that is stronger than the effect of perceived formal sanctions." (Akers 2000, p.23) The impact of formal sanctions on drug use has been the subject of much research, so the focus on this paper will be to understand the role that informal sanctions play in the rhetoric used in the decision of whether or not to use drugs.

Theoretical Concerns Based on the Underlying Assumption of Deviance

In applying rational choice theory and deterrence theory to drug use, one must make the underlying assumption that drug use is deviant, which some find troubling. The easiest place to see why this assumption is troubling is with regard to the application of informal sanctions. If the government is attempting to deter drug users from an action that is generally considered deviant, then informal sanctions will work in synergy with formal sanctions to deter individuals from drug use. If drug use, however, is not considered deviant, then deterrence must work without this synergistic effect. To establish whether drug use is considered deviant in the United States today, one must first answer the question of what exactly is deviance. Tittle states: "there is no generally agreed-upon conceptualization of deviance," but a necessary condition of such a conceptualization is that it "conveys the idea that deviant acts are inconsistent with the standards of acceptable conduct prevailing in a given social group." (Tittle 1995, p.124-26) The National Household Survey on Drug Use and Health, administered nationwide by the National Institute on Drug Abuse, reported in 2003 that 47.8% of the population had at some point in their lives ingested an illegal substance (National Household Survey on Drug Use and Health 2003). Such a finding would indicate that drug use is not considered deviant by a large portion of the population, although people may come to regard drug use as deviant even if they themselves have engaged in such activity. Or, it may be the case that drug use is generally considered deviant, but there are subcultures of the population in which it is not.

Becker argues against the uniform depiction of drug use as deviant. In his book, *Outsiders*, he sustains the claim that the relationship between the drug user and the establishment is one of reciprocal accusations that the other is an outsider. The establishment labels drug users as deviant, but users themselves label their behavior only as pleasurable. Becker believes that there is a fundamental difference, specifically in the case of marijuana users (though this could be applied more widely to drug users in general) between users who are secret deviants because their use is under the radar of authorities, and non-users. The project for Becker, then, is to separate the secret deviants from the publicly identified deviants, because he believes that their motivations and the effects of their deviance are significantly different, and therefore, that each of these two groups requires a different explanatory theory.

Becker also makes the intriguing point that even harm reductionists, thought to be more tolerant towards drug use, are just as entrenched in the concept that drug use is indeed deviant as others less tolerant toward drug use. To say that illegal drug use is a disease to be treated entails the presupposition that there exists an objective view of deviance, and that this objective deviance needs to be treated. It may very well be the case that drug abuse and alcoholism are objective diseases to be treated, but that does not imply drug *use* is always objectively deviant, nor to claim that deviant behavior always needs to be deterred. Whereas syphilis is a disease in any cultural context, deviance is culturally defined. By extension, drug use is not deviant in all cultural contexts (as is partially evidenced by the high percentage of Americans who have used illegal substances). "The medical metaphor . . . accepts the lay judgment of something as deviant and, by use of the analogy, locates its source within the individual, thus preventing us from seeing the judgment itself as a crucial part of the phenomenon." (Becker 1963, p.6) Becker wants to get to the bottom of why marijuana use is labeled as deviant, and whether it is considered deviant by all subgroups and subcultures, and he is right in acknowledging that the first step towards doing so is not thinking of drug use as a disease to be treated, but rather as a phenomenon of (possible) deviance in creation.

Part of the reason there may be a discrepancy in viewing drug use as deviant has to do with how drug users are prosecuted, based on their membership in certain groups. Becker makes the insightful remark that laws are not applied uniformly among different social classes, nor between blacks and whites, nor between men and women. As such, the middle class white drug user is likely to remain a secret deviant who does not label him or herself as such, whereas other users are more likely to become publicly known for their drug use and therefore suffer greater consequences. Becker reminds us that this distinction among the socio-economic classes does not happen just at sentencing, but at every stage of criminal behavior: the privileged as less likely to be arrested for drug crimes, less likely to be booked when they are arrested, less likely to be convicted when booked, and receive a lighter sentence when they are convicted. The Reagan Administration's move towards mandatory sentencing, which occurred after the publication of Becker's work, equalized sentencing across the board, but did not address the problem of offenders not uniformly entering into the penal system in the first place.

It seems clear that both rational choice, deterrence theory, and Becker's "Outsiders" theory require an understanding of whether drug use is considered generally deviant, and if so, how it comes to be considered as such. To put it another way, there needs to be an understanding of the process by which people decide whether to use drugs, and whether this process acknowledges the formal and informal sanctions associated with drug use, if the latter exist. Rational choice and deterrence theory require the underlying component of informal sanctions in order to fully function, which in turn requires that drug use be generally considered deviant. If those who would enforce informal sanctions do not consider drug use deviant, then they will not as readily reinforce an anti-drug use message. If Becker is correct (at least in the case of marijuana) it seems that as a society we have blurred the line between the extrinsic problem of drug use, that is, the outside sanctions imposed by policies influenced by deterrence theory and informal sanctions), and the intrinsic wrongs associated with the use (harms by the substances, etc.). It would seem, then, that people believe drug use to be wrong because others around them do so. Parents tell their children not to smoke marijuana because it is wrong, and when asked *why* it is wrong, answer that it could lead to incarceration or job loss, both of which are external to the potential users' moral structure, which influences their rational choice.

Methodology for the Analysis of the Online Rhetoric

How does one go about establishing whether drug use is either generally or subculturally considered deviant? One popular way to find out what people really believe about an issue is to see what they write about it online, where they are anonymous and there are relatively few consequences attached to what they write. The idea of being protected by the anonymity of online discourse is sometimes referred to as the "cloak of anonymity." This cloak protects online rhetoric in several ways: the possible negative feedback one would receive from another poster is also anonymous, and thus less likely to prevent the user from posting his true thoughts; the user herself is protected from informal or formal sanctions based on the rhetoric by its anonymous nature; and, any possible negative feedback on the rhetoric is delayed by the online nature of the rhetoric (which does not happen in "real time"). In order to capitalize on the cloak of anonymity and related effects, I decided to analyze online rhetoric taken from two websites on opposite ends of the drug use debate. Because drug use is illegal, and a sensitive and personal subject, I believe an analyzing online rhetoric will be particularly fruitful. The first website, TheAntiDrug.com, purports itself to be a resource for parents trying to stop their children from using drugs. The organization who runs this website has been prominent in national magazine and television advertising campaigns, including spots on CBS and *TIME* Magazine. The website, the major focus of the group, clearly takes the position that all illicit drug use is negative and that all such behavior should be deterred. The second website, Erowid.org, is renowned for taking an adamant harm-reductionist stance to drug use. The basic premise of such a stance is that people will use drugs, and the best thing to do is to provide them with the information necessary to make the use as safe as possible.

The structure of the interactive portions of these two websites is quite different. The parent website, TheAntiDrug.com, purports to be a resource for parents trying to spot and stop their children's drug use. Most of the web page is devoted to parental resources, such as summary of new research about the effects of drug use and information about the different drugs. Most of the information is a synthesis of information available from the National Institute on Drug Abuse. The website does provide an area called "Sharing and Listening," which is dedicated to allowing parents and children to share their experiences with drug use. Parents and children alike can post directly to this section of the website, but must do so after reading the following disclaimer:

> TheAntiDrug.com welcomes your submissions. Before being posted to the site, all submissions will be screened for topic relevance, and may be edited for space and to ensure that no personally identifiable information is displayed. Submissions may be used in other parts of this website.
>
> Advice and comments in this are posted by site advisors and have not been reviewed by drug prevention experts or endorsed by the National Youth Anti-Drug Media Campaign.

All the stories in the first section of the "Sharing and Listening" portion of the site were analyzed for this paper.

Erowid.org allows all users to post their experiences about their drug use, and structures the posts around the predominant drug associated with the experience. The goal of these stories is to provide first-hand accounts of what to avoid or what to pursue with regard to a specific substance, along with the individual users drug experience, in order to be able to gauge what taking this drug will be like for a potential user. From the "Experience Report Reviewing" guidelines on erowid.org:

> The design goal of the Experience Vaults is to act as a categorized repository for the long-term collection of people's experiences with both psychoactive substances and techniques, and to make those experiences easily available to people searching for information about reported use, effects, problems, and benefits. Our editorial goals are to weed out completely fraudulent entries and to keep the texts focused on the first person experiences of the authors.

Posters identify themselves by a "handle" (an Internet name), and are asked to describe what drug(s) they were using, how much they weigh, what dosages they ingested, and then to qualitatively assess their experiences. The posts are arranged by which drug was the predominant drug in the experience being addressed, though the format does allow posters to indicate if they were engaging in polydrug use.

Erowid.org has a plethora of posts discussing drug use, whereas TheAntiDrug.com had fewer than fifty. In order to have a similarly sized sample set, I used Erowid.org's internal search engine (powered by Google) to search the entire contents of the site. I searched for posts made within the last month that included normative words (such as right, wrong, moral, immoral, and should). I then read through all the instances, and established whether these instances were actually examples of normative claims, or if they were other uses of these normative words. Exclusions were based on the normative word not being used in a normative context, as in the statement "I placed the syringe to my right." After discarding the non-normative posts, I then had my sample set of Erowid.org posts. I chose to focus on the normative claims made on Erowid.org because I reasoned that posts including normative claims would discuss how people came to the conclusion of whether to use the drug in question, and then post about their experience.

Otherwise, I would have had to make arbitrary judgments about whether a post "counted" as discussing the decision process as to whether to use a drug or not. Because I am trying to capture how people make the decision of whether to use drugs, I thought that focusing on moral terms, I would obtain posts where people discussed the decision procedure by which they came to the conclusion to use drugs. Focusing on words such as "jail" or "illegal" or "prison term" would have led to posts discussing how users overcame their fear of formal sanctions, which was not the focus of this paper. Rather, by focusing on moral terms, I hoped to find posts wherein users discussed their rational decision to use drugs. Finally, I compared those posts to the deterrence rhetoric on the parent website.

What Users and Parents Had to Say

Surprisingly, the rhetoric from both websites seemed to be devoid of conversations about either moral considerations (how church standing might be impacted by drug use, Biblical arguments against or for drug use, utilitarian or Kantian grounds, etc), or of what is traditionally understood as informal sanctions (loss of standing within a community or family, exclusion from social groups, etc). Rather, both websites focused on the actual empirical effects of the drug use, or what I will later refer to as causal-empirical deterrence.

TheAntiDrug.com Website

There were several recurring themes in the rhetoric found on this website: users do not authentically choose to use drugs, and drug use leads to exclusively negative consequences, most often to death. In sharp contrast to the rhetoric on Erowid.org, users were depicted in this website as not *choosing* to use drugs. When a choice is made, it is blamed explicitly on the alleged friends of the user, or how the decision to use the drugs is stated as being unknown (and unknowable).

Exemplars of Discussion of "Choosing To Use Drugs"

Now keep in mind that my friend has never done any type of drug, smoked, nor drank. She went to this concert with 3 other friends. Her "friends" gave her a pill. I was sleeping at a friends house when i got a call at 6:00 in the morning saying my friend has been put on life support for taking the pill. In the evening at some time the next day, they pulled the plug. [theantidrug.com]

My daughter took ecstasy, unwillingly, on December 30, 2000. She died January 4, 2001. This was her first and only time taking the drug. I will live with this always, and so will her brother and the rest of our family. Ecstasy KILLS. [theantidrug.com]

Later I found out her friends bought and gave her oxycontin, a drug for cancer patients. Till this day I have no idea how she took the drug, why she took it but I do know they watched her die! Her friend said she tried to give her cpr and the others would not call for help They let her lay on the floor for 12 hours. [theantidrug.com]

Even those who post about their own past drug use do not see themselves as having made a rational choice in using drugs, but rather see themselves as having reacted against parental or familiar pressures which directly led them to use drugs. The major message in these posts is that parents and other family members are responsible for the drug use, and are specifically able to prevent it through paying more, and the right kind of, attention to their children. Even in this section, blame is put on family members, parents, and friends, but there is no discussion of *choosing* to use drugs or discussion of experimentation. Rather, drug use is viewed as a reaction against the lack of positive influences. The main answer to the problem seemed to be for parents to become more involved in their children's lives, either through more direct parental supervision or simply by paying more attention to the children.

Exemplars of Neglect Leading to Drug Use, or, "Pay Attention!!!"

Because of my addiction to marijuana I went from a 4.0 student to a 2.5 student and dropped out of all the sports because I became lazy and distracted. I first tried the substance believing that "you can't be addicted to pot." WRONG. I am a perfect example of the fact that you can, and IT IS the gateway to other, harder drugs. I am just now able to control this addiction and many others due to circumstances that I'd rather not share. The reason I write this is to show parents that sharing things with your kids really does matter. Most 13 year-olds aren't able to have a very strong willpower to saying "NO" without knowing facts and knowing how their parents feel about drugs. Who knows, if my parents would have talked to me about drugs I might be a totally different person and a whole lot happier. [antidrug.com]

I was 14 when I smoked my first joint. I liked the way it felt to be stoned. The fact that it made my parents mad was an added bonus. I was a middle child with average looks, abilities and so forth. I felt invisible in my family. Not really neglected, but the focus rarely fell on me. So I got to do pretty much whatever I wanted. That included staying out late with my so-called friends. I started using drugs more for recreation than anything else at first. Things changed the first time I got caught though. My parents hit the roof. And for the first time in my life I had a role in my family. I was the "bad" kid. Which, unfortunately, was OK with me. At least they were paying attention now. [theantidrug.com]

As for the discussion of the negative consequences arising from drug use, the posts did not generally distinguish between negative consequences that were directly related to the drug use, and those that were related to external formal sanctions attached to drug use. Furthermore, causal links between the drug use and the negative consequence could be tenuous at best and still considered acceptable for submission. For example, one poster said that his brother committed suicide and was a marijuana user, but the poster did not elaborate as to any other factors that may have been involved in his brother committing suicide:

"Steve's dead." Worst thing I ever heard. I wanted to be just like him. Sounds strange, but he was my role model. He was two years older than me, and I just knew he was the coolest person in the world. Well, Steve was caught with Pot about 2 years ago. It was a downward spiral after that. Last year, he shot himself in the head. He never asked for help. We all loved him. He always felt trapped. We should have reached out to him. I regret that. It's too late now. [theantidrug.com]

In the attempt to "scare kids straight" away from drugs, behavior is attributed to drug use, even if no necessary causal relationship could be established between behavior and drug use. When discussing their children's drug use, parents did not consider any other possibilities, such as an underlying causal factor leading both to the drug use and the other dangerous deviant behavior(s). Specifically, drug use is discussed as the sole cause of unhealthy, risky sexual practices, though the possibility of such behavior without ecstasy is not entertained:

> My daughter contracted HPV 18 and is now dying of adenocarcinoma of the cervic. She aparantly contracted it after her inhibitions were let down due to her ectasy use. Let your daughters know that one night of indescretion due to drug use could have permanant reprucusions [theantidrug.com]

This causal chain could have been broken at any time: the sex that led to the HPV did not necessarily occur while on ecstasy, regular pap-smears might have caught the cancer in time to save this young woman from dying, and we have no assurances the mother is not exaggerating the claim of the daughter's possibility of dying.

Possibly related mitigating factors associated with the tragic consequences are rarely considered, and when they are, they are quickly dismissed. In the earlier example of the young woman who died because her friends would not call for help, the possibility that it is the current criminal nature of drug use, and the formal sanctions associated with illegal drug use, which most likely kept friends from taking the victim to the hospital, is never entertained. It may have been that friends were afraid they would end up in legal trouble if they took someone to an emergency room who had clearly overdosed. As such, in order to protect themselves, they let this young woman stay on the floor and die instead of obtaining medical attention for her. If it is the case that friends did not bring the victim to the hospital for fear of formal sanctions, it is intriguing that formal sanctions were strong enough to deter users from helping someone instead of watching their friend die, but they were not a strong enough factor to deter them from using the drugs in the first place. The posting by the young woman's mother, however, definitely seems to indicate that the causal-empirical effects of the possibility of death would be enough to prevent someone else from taking ecstasy.

Another possibility in the above case is that the young woman's friends were too intoxicated to drive, but this does not seem likely if one of them was at least cognizant enough to attempt CPR. All of these other possibilities are not considered: this tale is only told in the context of "do not do drugs," whereas (as we shall soon see) on Erowid.org drug use is discussed, but in the context of "know how high a dosage you can handle" or "always make sure you try a drug for the first time in the company of people who will help you if you have a medical emergency." In contrast, one Erowid poster said he went to the precaution of having a registered nurse present when he first experimented with heroin.

One poster discusses the possibility of formal sanctions being more responsible than the potentially lethal property of a drug for her son's death, but quickly discounts it:

> Hypocrits and junkies would argue that legalization would have saved my son and his friends, because they wouldn't need to cook DTM in a kitchen, but that is just rubbish, the drugs would kill them anyway. Remember my son, and for your sake do not follow his tragic mistakes! [theantidrug.com]

Formal sanctions definitely play a part in the deterring discourse, although there is a sense in which they appear less successful than the rhetorical strength of talking about the direct negative effects of drug use:

> I had found some marijuana in my childs room several times but then one of my regular searchs I found a kilo of cocaine under his bed. I was somewhat shocked and of course my first response was to call the police. When the officers arived, in large numbers, they stormed my son's room and took him out by force as well as the thousands of dolars worth of cocaine. He is now in prison for ten years. [theantidrug.com]

> I just found that my teenage boy smokes pot. . . . I had a long talk with him, and explained to him the dangers of pot. How he could have his car taken from him for having a seed found in it, and all his cash seized from him for having even only one joint. I explained to him how being caught with marijuana, even only ONCE, can keep him from voting in any election ever again. I told him that if he is ever caught with weed, that he will be uneligible from receiving student aid from the government. I even showed him newspaper articles where people were killed by the police for having some weed. [theantidrug.com]

The following post best encapsulates the paradigm of the website, which is to highlight catastrophic causally empirical effects of the drug use, which is in turn done without the full choice of the person ingesting the substance:

> I went out to a "rave" party with a friend. [...] She called it an "atomic joint." I didn't know what that meant, but I was so out of my mind from the pot that I didn't question her. I took a long, deep hit of the joint and it in my lungs. The next thing I can remember with clarity is the florescent lights of the hospital hallway whizzing across my feild of vision while my girder was being rushed to the emergency room. We never were able to find out everything that was in that joint in addition to the marijuana and the doctors were left with their hands tied behind their backs. Now I can't move my arms or legs. I now have to spend the rest of my life in this filthy wheelchair. [theantidrug.com]

The Erowid.org Website

The discussion of the users on Erowid.org also centered on the physiological and societal ramifications of their drug use, but their reports did not reflect the tone of TheAntiDrug.com. Whereas the parents' website stressed the negatives of drug use, the posters on Erowid.org seemed to focus on wanting to capture the drug experience, almost as though it were a personal diary entry. In sharp contrast with TheAntiDrug.com, users described themselves as being in control of their drug experiences, having made a rational decision after much research into a substance in order to decide whether this was the right drug experience for them. They categorized their drug use as being something they *chose* to do, and when they talked about friends helping them to get the substance, it was in a context of being given the ability to take the drug, as opposed to have no choice in its consumption:

> The last several months, my interest in heroin has been piqued. So I gathered as much information about it as I could and have talked to several people (several bluelighters in fact . . . you know who you are, thanks for your help) who use heroin . . . those who use it occasionally without problem, and I even confronted two "junkies." These people have a severe dependancy on heroin, and have not been able to successfully quit. After weighing it out, I decided if the opportunity arose, I would accept. ["Silver," erowid.org]

A main focus of the discussion on Erowid was the ability of being in control of your drug use, instead of letting it control you, and how one goes about being in active control of their use. Many users discussed waiting to ingest a drug only when one felt ready to do so, and seemed to believe they were not likely to become addicted. There was not much discussion of how one is to avoid becoming an addict, though the key is apparently not to use a drug too often, that is, to keep the use sporadic. There is very little discussion of how one is supposed to achieve such a balance, but these users are aware of the fact that junkies exist, as is evidenced above in Silver's post. The focus is definitely on making sure one ingests a substance in the safest way possible. Later in the above post, Silver discusses all the precautions he took in his first heroin experience: he bought his own needle and syringe. He also had the heroin administered to him by the girl-friend of the friend who provided it to him, who was a registered nurse.

Users on Erowid did admit to having had outside psychological problems as causal factors in problems they may have experienced, but most felt these problems were concurrent or prior to the drug use, and that their psychological problems were not caused or exacerbated by the drugs, but rather were only made more evident:

> I've had major problems integrating the psychedelic expirience with my life, I think I was bound to have problems with anxiety/disassociation, I don't blame these personality traits on drugs; acid just made them more visible. ["Giger," erowid.org]

There is a definite sense here that users consider themselves to be making a rational choice to be using drugs, which would seem to fit in with the status quo theoretical framework. Criticisms of the drug framework resonate with Becker's claims of drug users feeling like outsiders to the system, as is highlighted in these posts. Some users went so far as to say that drug use is criminalized because of the positive effects of drugs:

> It is commonly stated that illegalizing drugs is the "moral" for a government to do, since drug use is thought by some to be immoral, even to degrade the moral fortitude of citizens. But the governments taking this "moral" stance mostly sanction and support the use of drugs like alcohol and nicotine, as do the vast majority of those citizens "morally" opposed to illicit drug use, the bulk of who themselves are drug users. [...] "Moral" condemnation by the majority of Americans of some substances and not others is little more than a transient prejudice in favor of some drugs and against others. [erowid.org]

> The ruling class is only trying to deter you from using drugs, with drug tests, anti-drug laws, and anti-drug propaganda, etc., so that you can be more easily controlled and, more literally, used for their purposes. [...] It is easier for those in power to control you if you are just like everyone else who cooperates with their systems and rules – systems and rules which are set up and administered to benefit THEM, not you or your mental health. [erowid.org]

> In retrospect, i can understand why drugs like MDMA and marijuana are illegal. I definitely do not agree with it, but drugs like these and LSD, peyote, etc. are mind expanding drugs, they free your mind to let you observe the bullshit our society puts its people through, the futility of capitalism, the corruption of government, drugs like these are illegal because the authorities are afraid of what could happen to their oppressive system if enough people actually understood the nature of american government and market capitalism. I have gained a new understanding from this experience, one that will never leave me. [erowid.org]

As critical as these posters are about being controlled by other forces, and are just as concerned about being controlled by the drugs in the same way. There is definitely a cost-benefit analysis based on pleasure (which reminds us of the utilitarian roots of rational choice theory). Pleasure is understood both in terms of its standard, hedonistic definition, as well as in terms of the pleasure of the mind-opening experiences. The rational choice is not between the pleasure and the formal and informal sanctions that might arise from using drugs, but rather between pleasure and the causal-empirical deterrence factor of the effects of using the drug itself. If formal sanctions are discussed at all, they are understood as yet another cog of the government machine, to be rejected by these posters. Posters ARE aware that these drugs can possibly enslave them in addiction, and that this would be submitting themselves to the same sort of control they are rejecting on a policy level. They think the pleasure outweighs the risk, although it is the users' responsibility to minimize that risk as much as possible:

really screw around with it, because hopefully by then you'll be directed and independant enough to integrate the experience into your life. [erowid.org]

It is not my intent to imply that the rhetoric on either website contains the correct approach to establishing a procedure by which individuals choose whether or not to use drugs. And, although the catastrophic effects discussed on TheAntiDrug.com seem out of touch with the experiences being posted on Erowid, one must keep in mind that users on Erowid are posting about their own experiences, and it is impossible to post online about one's own drug experiences which lead to death. Earth, one of the three administrators for Erowid, writes in personal email correspondance dated August 15, 2005: "I would note that the Erowid policy is that we only publish first hand reports EXCEPT in the case of serious injury or death for which there is any kind of documentation that the injury took place." On the other hand, a parent who has just had a child die is unlikely to know to go post to the site, whereas TheAntiDrug is more well-known in mainstream circles. So, TheAntiDrug's format of letting parents and siblings discuss family members' drug use does provide for reporting of catastrophic consequences that might not otherwise be as reported. However, Erowid's fact-checking editorial process probably makes their information more accurate.

Conclusion

To the extent that the rhetoric expressed on these websites are typical of drug users more broadly, then rational choice and deterrence theories are correct that users weigh the possible gain from drug use against the possible negative effects However, the negative effects are considered in terms of the causal-empirical deterrence intrinsic in the substance use itself and not in terms of formal and informal sanctions. Causal-empirical deterrence has to do with identifying the negative effects intrinsic to the use, and gauging those against the possible benefits of the use as the primary weights analyzed for rationale, in conjunction to formal and informal sanctions. That is to say, I define causal-empirical deterrence as weighing the negative effects intrinsic to the use as the primary negative against the possible benefits of use, in conjunction but not secondary to formal and informal sanctions. Why would drug users be more concerned with causal-empirical deterrence than with the more classical aspects of rational choice theory? This is perhaps because the drug user, as Becker suggests, has already self-identified as being an outsider to the group enforcing these sanctions, and in refusing to buy into the framework, he becomes immune to its deterring effect. The application of rational choice and deterrence theories to drug use is unique in that illegal drug use is one of the few arenas where it is possible that the direct consequences of the law being broken may be more of a deterrence than the formal or informal sanctions externally imposed upon drug users for their actions. For instance, if a person makes the rational choice to steal a car, it is unlikely that the act of stealing is going to lead to physical or emotional harmful consequences, outside of a government system wherein sanctions are imposed upon the thief, either formally or informally. But, in the instance of illegal drug use, this is the case: there are physical and emotional consequences to drug use one must weigh against the possible pleasure gained.

The implication of the above discussion is that in the case of drug use, a shift should be made away from formal and informal sanctions, and towards highlighting the causal-empirical deterrence factor. Although it is beyond the scope of this paper to fully explicate how this could be done, there are certain necessary requirements such an approach would require. First would be supporting accurate and unbiased research concerning the actual effects certain substances have on the users of these substances. Users on Erowid and TheAntiDrug describe widely differing drug experiences, which suggests that environment and precautions can widely shape how a drug affects the user. If causal-empirical deterrence is going to be used effectively, there has to be an unbiased understanding of the likely effects of a drug, taking into consideration the environmental effects and how those shape the experience. Secondly, causal-empirical deterrence has the same limitation that classical deterrence has, namely, the higher the likelihood of the negative effect, the lesser the requirement for the gravity of the effect. TheAntiDrug's focus on catastrophic consequences of drug use, consequences which are challenged by the much more positive experiences described on the Erowid website, are probably an attempt to highlight drastic negative effects of drug use. Perhaps it would be more efficacious to focus on smaller but more certain negative impacts intrinsic in the drug use in order to more effectively deter the use of illegal drugs. In the long run, it is important that the negative effects of illegal drug use be presented honestly. Otherwise, the negative impact argument against the use of drugs will be discounted, and this argument will lose its effectiveness as a deterrent to illegal drug use.

Notes

1. I would like to thank Professor Edward Muguia for his feedback throughout the life of this project, as well as his continued faith and the opportunity to be involved in this book. I would also like to thank (in alphabetical order) Robert Garmong, Josh Perlman, and Megan York for their invaluable help with the editing of this chapter. Finally, I would like to thank Earth, one of the three administrators of Erowid.org, for the kind response and commentary. I would also like to thank Professor Deano Pape of Ripon College for the class I took with him on Computer Mediated Communication, which facilitated much of the research in this paper.

References

Akers, Ronald L. *Criminological Theories: Introduction, Evaluation, and Application*, Third Ed. Roxbury Publishing Company, Los Angeles, 2000.

Erowid: www.erowid.org. (Accessed April 2003)

Goode, Erich. *The Drug Legalization Debate*, ed. James Inciardi. 1999.

National Institute on Drug Abuse. "National Survey of Drug Use and Health." <http://oas.samhsa.gov/nhsda/2k2nsduh/html/toc.htm#TopOfPage> (2003).

Parents. The Anti-Drug. Selections were taken from the "Parents Sharing and Listening" portion of the website, which was available at: http://www.theantidrug.com/sharing_listening. The submittal form can be found at: <http://www.theantidrug.com /sharing_listening/parent.submit.html> (Accessed April 2003).

Tittle, Charles R. *Control Balance: Toward a General Theory of Deviance*. Westview. 1995.

Chapter Seven

Assessing the Likelihood of Internet Information-Seeking Leading to Offline Drug Use by Youth
Sarah N. Gatson

Some Assumptions and a Question about Online Communities

- All Internet is media
- All Internet is communication-information
- All Internet is networked
- Much of the Internet is two—(multi-user) way communication
- How much of that multi-user communication is community?

Definitions of Community

Our commonsense definitions of community involve *locality* and *common government*, as well as *common interests*. Our sociological definitions incorporate the commonsense, and classically further demarcate kinds of community, with Töinnes's *gemeinschaft* incorporating the idea of local, traditional, face-to-face, and close emotional ties, and his *geselleschaft* being the passage from gemeinschaft to the looser, associational and impersonal ties of the industrial age

(1887). Often, discussions of geselleschaft incorporate laments for the demise of gemeinschaft—the loss of community in the ever-increasing pace of the technological advance of humanity. This is an already well-worn complaint about the development of the Internet.

At the same time, when we discuss the promise of the Internet, we often find ourselves focusing upon how Computer-Mediated Communication (CMC) *can* rapidly make gemeinschaft from a geselleschaft situation. Rheingold's (1993 [2000]) discussion of the field of possibility and actual accomplishments of community shows how and where folks are situated in relation to their computers, and to others scattered across the network of hardware, software, and users (see also Wellman, Salaff, Dimitrova, Garton, Gulia, and Haythornwaite, 1996; Wellman and Gulia, 1999; Wellman, 1999; Putnam, 2000: 148-180; Malley, 2003). There has been a wide range of interaction that has occurred historically and contemporaneously in CMC, and this ranges from purposeful (intentional) community in which members' lives become densely interconnected, to emergent and amorphous community that may involve some interconnection, to very loosely linked and dispersed networks and interactions, what we might characterize as *scenes* rather than communities (Berger, c2001; see Witte, Amoroso and Howard, 2000; Howard, Rainie, and Jones, 2001; Georgia Tech Internet Survey, 2000; Shirky, 2002; Malley, 2003).

Again, one of the reasons why concerns over the Internet as a communications technology have surfaced is its potential for influencing the development of human communities. Clearly, concerns over this new technology and concerns over other uncontrolled/uncontrollable cultural products and practices—namely drugs, drug usage, and the rave scene—have dovetailed to the point where the similarities among these subcultural arenas need analyses. Rheingold notes that unintended consequences in CMCs are of major concern, "A continuing theme throughout the history of CMC is the way people adapt technologies designed for one purpose to suit their own, very different, communication needs" (1993:7). Persons interacting in online contexts face a dilemma. On the one hand, they are presented with a new context for mediated interaction where there has been an aggressive presentation that the rules have changed, and things are up for grabs. At the same time, declarations of newness are (necessarily) rooted in extant cultural discourse (see Gatson and Zweerink, 2000: 106-7; 125-26; Gatson and Zweerink, 2004: 25-27).

Deviance and Community

The consequence of the media (information) function of the Internet is that it is unlike the traditional offline scenario of an individual exercising her power/right to information-seeking. One might go to a bookstore (local and independent or large and chain/national), or library, peruse the shelves, and select a copy of Zimmer and Morgan's *Marijuana Myths, Marijuana Facts: A Review Of The*

Scientific Evidence, the McKennas' *The Invisible Landscape: Mind, Hallucinogens and the I Ching*, Pinchbeck's *Breaking Open the Head: A Psychedelic Journey into the Heart of Contemporary Shamanism*, or Barnaby's *Absinthe: History in a Bottle*, and learn about the politics of drug use, the spiritual argument for drug use, and perhaps even how to use drugs, or use them more effectively. The information is available, certainly, but how costly, in terms of the likelihood to be tracked in one's purchase/borrowing, in terms of money and time, and other resources?

The newer online scenario could involve sitting down at a computer—one that is more or less anonymous (e.g., one's own, dedicated desktop [non-portable], to one's laptop and remote access, to a public site such as a library or café)—and linking to (for little to no cost) the same type of information potentially one hundred-fold. Depending on the length of one's surfing session, reading comprehension, time and other non-monetary capital resources (such as non-interfering personal relationships, unlimited access at any given time of day, etc.), and general stamina for computer-mediated communication, one could not only link to sources of *information*, but to a vast network of *persons*.

The assumption here is of course that the latter scenario involves being more unobserved, more uncontrolled, and being more easily drawn into either a "virtual" life (in a derogatory sense), and/or a more dangerous, if not thoroughly criminal, set of situations and practices. However, as Lessig and others have clearly shown, the surveillance potential in the Internet environment has been well developed, and will likely continue towards more surveillance and control rather than less, even while netizens themselves seek to self-regulate and debate and practice the boundaries of anonymity and community (Lessig, 1999; see also Demac and Downing, 1995; Gandy, 1995; Schiller, 1995; Sclove, 1995; Mayer-Schoenberger and Foster, 1997; Goldring, 1997; Gurak, 1997; Reidenberg, 1997; Wallace and Mangen, 1997; Smith, McLaughlin, and Osbourne, 1998; Godwin, 1998; Sobel, 1999; Loader and Thomas, 2000; Demac and Sung, 2000; Gatson and Zweerink, 2000: 126-27; Lipschultz, 2000).

The social problem of the nexus of new technology, youth, and drugs, is one that parents and other authority figures define. One major approach to the problem combines a concern with deviance and the accuracy of information that may or may not be put into practice. Stemming from this is a concern over whether access to information will (necessarily) lead to particular practices/behaviors (drug-taking), or indeed, to subcultural immersion (an online/offline merger of communities involved in raves and drug-taking, understood as fundamentally linked activities). In statements regarding the effect of communal ties upon deviant (read as socially defined as undesirable *and* outside the norm) behavior, both Emile Durkheim and Robert Merton and their use of the concept of *anomie* (in conflict with societal norms; without law) highlight the importance of the question, to which communities are we attached (Durkheim, 1951; Merton, 1938; see also Hirschi, 1969)? As I noted above, we should pay attention to the *kinds* of communities to which youth (and others) are attached online (and offline). In other words, what are the levels of intensity, and just exactly what kinds of interactions are people developing through participation in CMC?

In the larger project from which this chapter emerges, we presented a range from essentially close-knit communities (Lessem, 2003; Kotarba, 2003; Gibson, 2003; Murguía, 2003), to instrumental communities of harm reduction (Gibson, 2003; Murguía, 2003; Gatson, Chapter 2, this volume), and finally to dispersed networks with crucial nodes that represent significant chunks of the rave/technology/drug matrix/nexus/scene (Gatson, 2003; 2004a). It is this final type that I deal with here. The methodology for this project involved the extension and adaptation of traditional ethnographic methods to the practice of online-ethnography. Because the subculture and networks surrounding the use of the so-called "club drugs" is one in which "underground" and countercultural, if not always illegal, activities abound, and because the subculture includes a significant technological component, entrance into the subculture, and the necessary building of trust in the research population sample was admirably served by beginning with the Internet arenas of this group. There was a wide variety of sites we observed, ranging from commodity sites, reading/information sites, networked sites, and what we dub "true online communities" (ongoing, stable participation, stable URLs, interactive CMC).

Overall, this is a "scene" or a movement—an amorphous subculture—that is indeed interlinked online and offline. Only some of these sites/groups/arenas, amount to a community—well-defined and -defended spaces/places/identities. To be a raver—is this to be a member of a community, or merely a participant in a cultural scene? The answer to that question is going to be very different for the intense and regular participants in some online places, or regulars in the offline rave scenes in various localities, as compared with folks surfing the 'net for a specific piece of information (see Lessem). Can we say the same thing about the drug users? Where do the communities overlap?

The Discourse of Drugs and the Rave Movement

In the subculture under investigation here, there is a merging of several both transgressive and "progressive" (in the sense of the development of new and adaptive cultural practices) concepts and practices. First, there have been two uses of technology important to this subculture: a) the use of "techno" music and video production, and b) the incorporation of new technologies and media in communication about the community. Second, this technological savvy and affinity has a further importance in that the drugs of choice in the subculture (MDMA, GHB, etc.) are lab-produced. Finally, an informal, and aggressively anti-corporate, "do-it-yourself-and-share" value system is connected to these technological and chemical practices/values. These values, networks, and systems overlap and merge with other countercultures and underground cultures, such as "hacker" culture, and "ModPrims." Mizrach notes, "[these] subcultural groups [of hackers, ravers, and modern primitives] . . . unite culture, technology, and communication." As self-described members of the rave culture have put it:

> The actual concept of raves is not new—it is as old as time itself. As the base level, raves are very comparable to American Indian religious ceremonies, i.e., pow-wows, and also to the concept of the Shaman in Eskimo and Siberian society—where music is the key towards pulling oneself into a unique emotional and psychological state, a state in which one experiences washes of sensations and visions, not delusions, but visions. Sounds very hokey in print, but I'm sure MANY of you out there know what I'm talking about. The hypnotizing effect of techno music coupled with the seamless transitions and thematic progressions of rave DJs as the night progresses can be QUITE intoxicating, resulting in what could be closely compared to a religious experience. Music in general has always been able to sweep people off their feet, but what distinguishes raves are the concept of the _shared_ experience; a feeling of unity often arises, and people are open and friendly to one another. There is a loss of that "attitude" that is omnipresent in normal clubs and even in life in general. People are celebrated for who they are, not what they aren't. (Brian Behlendorf, 1994)

> "It's all about mixing things to get new hybrids." (From *Modulations*, film, 1998 Caipirinha Productions)

Thus, the sampling and lack of traditional musical instruments of the rave DJ, the cobbled together (and often illegal) use of public space, visual and sound media, the expectation of transported consciousness—through both the natural and drug-assisted high—and the feel or "vibe" of the crowd exist in both cyberspace and non-virtual spaces for the intended subjects of this research. The anonymity *and* potential space for meaningful connections are present in both arenas of this subculture's spaces of interaction.

Methods at Erowid

I focus on a spatially dispersed network of people and websites involved in a discussion of music, technology, and drug use. Specifically, information and interactive websites surrounding the rave/dance/drug-using scene, housed on a website arguably central to this scene. Beginning in September 2001 and going through June 2003, I engaged in qualitative assessment and observation of the networked counter-cultural website known as Erowid. This site is a major clearinghouse of information as well as a densely interlinked node of the overlapping concerns of music, art, spirituality, psychoactive substance use, and political-legal discourse. Erowid is a typical representative of an important, heavily trafficked and well-connected *node* in this *scene/network*.

The stated purpose of Erowid is "documenting the complex relationship between humans and psychoactives." The site's owners describe it as

> an online library of information about psychoactive plants and chemicals and related topics. The information on the site is a compilation of the experiences,

words, and efforts of hundreds of individuals including users, parents, health professionals, doctors, therapists, chemists, researchers, teachers, and lawyers. . . . The information found on the site spans the spectrum from solid peer reviewed research to fanciful creative writing.

The site is an important one in this scene, given that "over 18,000 people visit the site each day, making more than 2.8 million unique visitors in the past year." Erowid places itself first and foremost at the center of a national (international) socio-cultural and legal debate about the presence, use, abuse, and legitimacy of drugs.

> Erowid itself is a small non-commercial organization that has operated for more than 5 years in the controversial and politically challenging niche of trying to provide accurate, specific, and responsible information about how psychoactives are used in the United States and around the world. . . . Although the risks and problems are widely discussed, it's also clear that psychoactive plants and chemicals have played a positive role in many people's lives. As our culture struggles with integrating the increasing variety and availability of these substances into its political and social structures, new educational models are clearly needed. Erowid is founded on the belief that a healthy relationship with psychoactives is one grounded in balance, where use is part of an active, intellectual, physical, and spiritual life.

I began my assessment of Erowid and its own stated place in this debate by exploring its graphical and interactive aspects. More than a traditional clearinghouse, Erowid exists as a node in a perpetually linked-up network. The site offers a holistic ideology, whose major underpinning is free association and free exchange of information and ideas. Direct experience reports (with art, politics, and spirituality, as well as drugs) and information-immersion on these topics are the formats offered herein. The hosts of Erowid divide the site into four major areas of information/discourse: "Plants and Drugs;" "Mind and Spirit;" "Freedom and Law;" and "Arts and Culture." These areas overlap to a significant degree. For example, "religious freedom" as a subject has a presence on each page. The operators of this site then, if not all of its audience, do not compartmentalize the discourse of these issues.

As the major research question of the research team is whether and/or how the CMC technologies facilitate and proliferate the drug use of youth, it is not enough to demonstrate the overwhelming availability of discourse on drug use, including the sale of drugs, how to use drugs, what to do if your use results in ill effects, etc. E.g., everyone gets porn-spam in their email inboxes; but does everyone consume such material? Once the setting is known and outlined, we must then follow the links ourselves, and assess the actual ease of moving through this scene, as well as what people are doing with and in this informational network.[1] Thus, I next moved on to the "Reciprocal links of Erowid" section of the site.

Erowid itself does not offer computer-mediated communication tools that directly facilitate dialogic and other conversation-exchange opportunities. While the site offers the opportunity to submit personal reports, and to suggest the inclusion of material written by others, it does not have a chat room, BBS, or other live communication space. Although Erowid says, "Entries added to the guestbook are displayed live for all visitors to view," the guestbook is rarely used by users for responding to one another in a continuous conversational manner. The only clear responses to posts are from Erowid themselves. On the first guestbook page from 1997, Erowid states, "Happily, our guestbook entries don't scroll off the list. We at Erowid are dedicated to providing a forum for all information about entheogens and spirituality . . . not just those that are pleasant. :)." This indicates that the guestbook is maintained as a historical record which readers/members are invited to interact with, and although there are some embedded email addresses and websites, there is no dedicated space to engage in member-to-member contact on the site itself. Indeed, in 1998, one post indicated a desire for such a space, calling attention to the indication that savvy users distinguish between types of communication spaces,

> *A dISCUSSION FORM* [sic] *WOULD BE VERY NICE ALTHOUGH I'M SURE THIS WOULD BE VERY EXPENSIVE. MAYBE YOU COULD ASK FOR MEMBERSHIP FEES TO SUPPORT A FORM.*
> Robert Majors [embedded email address]
> —Tuesday, June 30, 1998 at 13:03:41 (PDT)

Further, Erowid does edit the guestbook, and so although live response is the format,

> **Ok . . . you asked for it . . . here's the Erowid guestbook censorship policy.**
> - Duplicate entries are deleted.
> - Entries by 14 yr olds swearing to hear their own voice are deleted.
> - Entries suggesting a correction are deleted when the correction is made.
> - Entries (or parts of entries) asking for information on buying and selling illegal or grey market substances will be deleted.
> - Information we believe to be innacurate will not be deleted, but will be commented on by us . . . in the name of safety.
> - Extremely confusing entries, or extremely confusing typos will sometimes have corrections or layout changes made in the name of clarity.
> - Changes requested in email by the author of the entry will be made EROWID [embedded email address] (from June 1997)

A Note on Ethics and Presentation of the Data

The mission of Erowid is to be a center of public information-gathering and—seeking. Since Erowid is itself a clearinghouse, much of the information it hosts is not self-produced. Thus, I have not disguised its name or online location. At the same time, many of the links housed at the Reciprocal page are not themselves quite so unambiguously public. The reciprocity function of this page—people voluntarily link themselves to Erowid—presents us with an embedded implication of public access. However, I was guided by the content and statements on the sites themselves in choosing what to reveal and what not to reveal about site location and those involved with the sites.[2] These differences become clear in the discussion of the data below.

Table 7.1. Data

Codes	Number (% Values)
Education Re Drugs	73 (18%)
Law and Politics (including Drugs)	67 (18%)
Drug Basics	45 (11%)
Entertainment (Music, Etc.)	96 (24%)
Religion/Spirituality	27 (7%)
Not Active	63 (16%)
Total	402

* Total number of active sites: a low of 282 and a high of 297 out of the 360 listed on the links page; %s do not total 100% because code categories are not exclusive.

Table 7.1 above details the general kinds of discourse and information offered by the Reciprocal Links page of the Erowid site. I coded first based on what discursive information was available from the links page itself, and perused those sites whose links were actually active on the days I visited them. Sixty-five of the Erowid links contained some form of significant CMC, including options to join mailing lists or listservs, chat rooms, or some form of BBS (a bulletin or threaded board), and links to Usenet groups such as alt.drugs.culture. Thirteen of these sites were no longer active or I received error messages when attempting to engage with them and/or their CMC venues.[3] Of these sites, fifty-two were still active as of April 24, 2003. In order to assess the likelihood of a person picking up a link from Erowid, and following it to the behavior in question herein—drug use—I followed the links, and read each of these sites in depth. I concentrated on the actual use of the CMC areas on each site, as well as any history of traffic, and an assessment of the existence of an in-depth community.

The protocol I created and followed consisted of the following steps: For all CMC-coded sites assess the likelihood of (a) following online informational reading with online contact; (b) following online contact with offline contact; and (c) the likelihood of the opportunity of following offline contact with drug use:

Table 7.2. CMC-Code Sites and Contact Likelihoods

Protocol	Contact Likelihoods						
	-3	-2	-1	0	+1	+2	+3
Online Information Seeking Online Contact (A)	2			4	2	5	39
Online Information Seeking Offline Contact (B)	3	2	2	10	8	14	13
Contact Likelihood of Drug use (C)	5	1	2	16	8	4	16

N = 52 sites

Elsewhere in this volume an analysis of the user/audience provided trip reports is made (see Crispino, Chapter 6, this volume). There are many more embedded email addresses on the early guestbook entries. This may indicate that Erowid has been used less and less as a communal contact portal. I would thus code Erowid itself as A=3, B=0, and C=1. In my assessment, full-blown information-seeking-leading-to-behavior (the +1 to +3 end of the scale) is a cause for concern in about 54% of these fifty-two sites. [4]

It is necessary here to discuss briefly what I mean by "concern." In those sites assessed at the highest end of the scale above, most had well-developed communities. In assessing the likelihood of arranging an offline contact in which drugs would be exchanged, shared, and/or sold, we have to pay attention to the density of ties within the online-based community in which a youth makes her/his initial information-seeking and online contact. Although there are interactions within these communities in which members discuss meeting up, and meeting up specifically for drug use, there are also discussions wherein they signal their wariness to do so, often specifically highlighting the suspected surveillance of professionals in the legal arena. Thus, a parent concerned with her/his child's potential for engaging in problematic behaviors—whether that be getting involved in online communities at all, or using that interaction as a gate way to drug use—would have to surveil online usage.

However, given my underlying assumption that an extant interest in the topic—if not in the actual usage—of drugs must be present for a person to go online seeking such information in the first place, it is somewhat problematic for the existence of these websites to be our primary cause for concern. In other words, online information and communal activities did not, and perhaps do not, instigate generalized youth interest in drug use. Online arenas may indeed *facilitate* further interest and indeed actual use, primarily by presenting the online entrant with a significant source of support by an actual or potential reference group. For example, if we compare the number of replies to posts versus views of posts, we can assess the likelihood that people read, but don't contact, the posters. The ratio just on the first page of the marijuana forum at Hippyland (Case #2) on the most viewed thread is 3 posts for every 112 views; other threads come in at 1 for every 11; 1 for every 12, etc. Reading isn't initiating conversational contact with a person, but may represent discursive contact with a reference group. It may also merely be occasional reading, a negative assessment of that which is read, as well as it could be reading-as-directive.

Examples of Site Discourse/Information, and Contact Likelihoods

I began the protocol with two sites that I had determined had both well-used CMC, as well as information on local gatherings and/or gathering spots. I emailed the initial webmaster contacts listed on the two sites,

> I found your site from the Reciprocal links page at Erowid. I've been studying Internet communities since 1998, and the network of drug discourse fascinates me . . . I've been reading your site as part of a research project I'm working on about Youth, Technology, and Drug Use. Although the project is funded by NIH/NIDA, I am not interested in being a narc. My part of the project specifically deals with the ease of moving from Internet environments to face to face contact and community formation. [I then included a brief description of the research, and the abstract of Gatson 2003] . . . I am mainly interested in learning how much offline contact you get with posters to your various forums, email, etc. Do you think there has been significant community development online and offline stemming from interactions initiated at your website?

I then planned to participant observe online, and to follow-up with any face-to-face community interaction that was available. In both cases, I was met with courteous and informative responses, but not the easy follow-up in offline contact that my initial perusal of the sites—nor my own extensive online experiences—had led me to expect.

Site #17: 4/21/03
Hi Sarah,

You wrote to me asking permission to join the [site name here] board. How ever, i'm not the owner the [site name here] forums; i merely link them to my website.

It's easy to join the [site name here] forums. All you need is a user-name/password entry. [site name here] is a techno-music community forum site. My site is "in-the-making." I'm working on creating a database of cultivation information about various ethnobotanicals. However, my site is still small at this time. So feel free to contact the owner of [site name here], if you wish; however that's not me :)

I took the initial contact's advice, and wrote to the site owner. I received the following response,

4/22/03
Greetings.
Hmmm, well it certainly is highly interesting that our small community was somehow found (through Erowid? That's mighty strange) but I'm afraid that we might not be a particularly good study . . . drug usage is usually below the radar, and the focus is mainly on a style of music which is seldom heard anywhere on this continent. Furthermore, my guess is that the privacy-oriented netizens of my community would not take kindly to being examined, though I personally agree in the abstract with any intent to research, learn, or otherwise contribute to humanity's knowledge-base. I suppose the other key ingredient is *activity* and we don't have very much of that—compared to many communities ours is VERY small. I've been involved with online communities since the early 90s, both as a participant and sometimes as a SysOp, and I recognize that a "critical threshold" usually needs to be passed for a community to truly coalesce. Ours certainly hasn't hit any such threshold—we're marginally eeking out a slumber-some existence, which suits us all quite well. Maybe we're internet burn-outs :) for the practical purposes of your study, there is little offline interaction initiated by our community I would guess . . . I've got some friends that come on, and eventually anyone into this music will meet everyone else, but the board hardly makes a huge impact here. To explain something which you may have mis-interpreted . . . our music is usually known as "psychedelic trance," which one may identify with illegal drug usage . . . but in fact, it is a form of electronic dance music well-suited for listening WITHOUT drugs. Though, of course, many people do drugs and listen to this music (as they do to any other kind of music). Naturally this is just my perspective . . . Well anyhow, I'm always open to hearing more about such projects . . . perhaps through some correspondence our different worldviews can overlap and communication can be acheived.

I had already registered with this site's BBS, but in deference to the webmaster's desire that I NOT participate, I only looked at the threads to assess the gatherings, and how easy it would have been for me (or, potentially, a youth with Internet access) to get to a gathering. I coded this site as A=3; B=1; C=-1 (See Table II above).

The next site I assessed was closer to me geographically, and seemed to have a great deal of emphasis on local gathering, and was the first site to delve into anything like political commentary on the legalization of drugs, marijuana in particular. I emailed the site owner as I had done with the first case and received the following in reply,

Site #18: 4/21/03
Hello miss Gatson, I agree with what you say about our youth seeing material like this could be negative. I have been seriously contemplating the fact that maybe i should remove the website. I started smoking at a young age and even though it is politically wrong i dont see marijuana as a dangerous drug. The reason my sight is up is because in 1984 i got arrested for a small amount of marijuana and it changed my life forever. It cost me an arm and leg in attorney fee's and court cost besides the fact i was looked upon as a serios threat to the community. Matter of fact the same community you live in. Thats right i was raised in BCS and went through the college station school district. **As far as meeting people who respond in my msg boards,forums etc. I havent ever met one person face to face. It something i wont do.** I personally think if the youth of today smoked marijuana and strayed far from the psychadelic drugs, LSD, X, magic mushrooms, speed and others they might have a chance at living to see a old age. Im sorry this is the way i feel but it is. I dont do hard drugs and never have but i have seen what they do to others. I smoke because i dont like alchohol. I tried drinking at the age of 17 and didnt like it. Marijuana is very mild and i am always in total control and aware of my surroundings. I think i am a very good person. Very compassionate about life and feelings toward others.

In the first case (#17), the community turned out to be linked from a working-site that was currently in the process of building to offer information about the Underground scene in a large Canadian metropolitan area. While long-established and in-depth, this community was not one seeking to advertise, grow, or be open to new members particularly. The information members posted about parties or gatherings was minimal, and a working knowledge of the offline locale for these folks would be necessary to get close to meeting them offline. A drug-induced state, as stated by the contact, was not the preferred method for this community's consumption of their music.

In the second case (#18), while the webmaster hosted a BBS and chat room, he himself was not interested in meeting people offline that he came to have contact with online (A=3; B=1; C=0). This site did link directly to Ya-hooka.com's BBS, as well as several other BBS and Chat sites that would very likely have a higher rating on this scale. There was also a mailing list option.

After these two initial contacts, I went through the rest of my list of these CMC websites/portals. The first of these is known as Big Fun. For the purposes of this project, its primary reason for inclusion has to do with its information on DXM (Dextromethorphan), particularly that found in cough syrup. This website vividly and non-anonymously describes a local geographical subculture, as well as having extensive, detailed, and well-developed CMC resources. I coded it as A=3; B=2; and C=2. This site contains excellent, qualitative network analysis, both geographical and interpersonal, by The Gus, the creator of the site. The site itself is a series of documentations of a sort of commune that was located in Charlottesville, NC in the mid-1990s.

This almost anthropological site links to other areas of the ongoing life of The Gus. He has since maintained a page with a series of journals documenting his post-Big Fun life, his move to San Diego (through 1998), and currently maintains a journal page documenting his life as a "35 year old married white male living in Hurley, New York, at the foot of the Catskill Mountains on the less irritating side of the Hudson River." This livejournal/weblog is current, with the latest entry in my protocol at April 22, 2003.

Although the Big Fun site itself is a historical documentary of life in Big Fun/Charlottesville, VA in the mid-1990s, it is likely that someone could at the very least take a "Big Fun local history" tour based on the information contained there. Online contact with at least The Gus, and others who maintain we-blogs/online journals would be extremely easy. Offline contact, depending on one's level of dedication to posting and participating in any blog community, could conceivably go on from there. Considering that the major drug of choice (although several are discussed, ranging from alcohol to prescription medica-tions) is an everyday, easily obtainable, legal product (DXM in Robitussin), no one would *have* to go online to either find information about it, nor to learn about it. "Drinking cough syrup" has a long history in substance use practices, from patent medicines, and the prevalence of laudanum in the nineteenth century, to the present day.

There are several "trip reports" for a variety of substances, including Ritalin and Nutmeg, with Dextromethorphan/Tussin seeming to be the most popular, with the most subcultural discourse attached to it,

> Big Fun began its life as the home of the Malvern Girls, three young adults from the Philadelphia Suburbs. From its initial disastrous housewarming and continuing on through the worst winter on record and into a spring infested with ticks, flies and cicadas, Big Fun, a stately yellow farmhouse in rural Cen-tral Virginia, saw many interesting sights: impromptu punk rock concerts, Dex-tromethorphan chug-fests, Nomadic Festivals, nazi skinheads, and (most im-portantly) record alcohol consumption. Like most tight social groups, Big Fun had its own language, history and collection of in-jokes. This website is de-signed to grant you access to the inner workings of this remarkable youth cul-ture as it actually functioned in the mid-90s.

> **tuss** (v)—to experience the effects of ingesting at least four fluid ounces of Tussin DM. **tussin** (n)—generic cough syrup. Robitussin is the far more expen-sive non-generic alternative if you feel you have too much money in your wal-let and you want to help a large corporation make record profits. If one wishes to tuss, not just any tussin will do; you need Tussin DM, because it is the Dex-tromethorphan Hydrobromide that makes drinking Tussin fun. Furthermore, Tussin used for the purposes of tussing should never have acetaminophine in it. A standard tussing dose of Tussin DM contains 10 mg of Dextromethorphan per little plastic cup, but you don't just do one of those pathetic little cups; you drink at least four ounces. Six ounces is better. Buy those big 12 oz bottles— they contain the most bang for the buck. Most people are usually able to drink Tussin straight without puking, even though the drinking is a miserable experi-

ence. Matthew Hart, however, does not appear to be able to keep Tussin in his system for longer than about twenty minutes. Farrell has perfected a sort of cocktail that contains Tussin DM, tonic water, lemon, and perhaps some kind of alcoholic beverage. It is marginally to acceptably drinkable, but one must drink a lot in order to tuss. **tussin city**—after one sits immobile for awhile while tussing, it is easy to forget how profoundly one has been affected by the Dextromethorphan. But if one then rises to his feet and walks around, he will quickly be reminded of the tussin. The process of being suddenly so reminded is termed "tussin city." **tussin euphoria**—drinking Tussin is such a terrible experience, there must be some reason the degenerate youth of today do it. That reason is tussin euphoria. Tussin euphoria is a mental condition where everything seems to be in a perfect state of harmony. All your friends are cool and all your new acquaintances you feel to be friends. You crave nothing, not sex, not food, not alcohol, perhaps not cigarettes (but I don't know, I don't smoke). And of course, you *never* have an urge to cough. **Tussin Tree, the** (1)—a little juniper tree that the Gus and Jessika gathered outside Big Fun prior to Christmas, 1995. The Gus decorated it with Tussin bottles and little cut outs from Tussin packaging (there were many such things in Big Fun at the time). The tree sat in a big three litre Cribari vino bottle upon which the Gus placed a blown up Food Lion Tussin label he had drawn with coloured pens. A little actual Tussin DM was actually added to the tree's water so it could tuss as it slowly died over the ensuing months. **Tussin Tree, the** (2)—on New Year's Eve, prior to 1996, it was a fairly warm night, and the Gus, on Tussin at the time, sat under a great oak tree just behind Big Fun and felt so good that tears came to his eyes. That was the night he met Wonder Boy Neek. Maybe it was the Tussin, but the Gus hit it off well with both Wonder Boy Neek and the tree he was seated under, the latter of which he called "the Tussin Tree."

Dextromethorphan Hydrobromide—the active substance in Tussin DM and a number of other cough remedies, similar both to LSD and PCP in structure. When taken in sufficient quantity, Dextromethorphan Hydrobromide brings about tussin euphoria. At right is the molecular structure of Dextromethorphan. (From the Big Fun Glossary by The Gus)

A far less idiosyncratic site that is directly about drugs per se is the Cannibis.com site (A=3; B=2; C=3), which hosts message boards, and offers some evidence of offline meeting. The interesting thing about the discussions I observed therein is how linked they were to gathering for political purposes,

> **Message #268** posted by corruptedyouth. February 08, 2003 03:42:21 ET
> I was talking to my friend about this idea last night, before I read the post about the organized smoking protest. We both thought this would be an EXCELLENT idea. I have some improvements to make . . . Most of the rallies should take place in Washington, D.C., for obvious reasons, for example on the steps of Congress. I understand that not everyone would be able to miss work (work?) and go to D.C. I like the idea about the parks, but if we were to get some SERIOUS publicity, and that is what we really need, we are going to need a MASSIVE gathering. Some other improvements I have been pondering are the use of more sy[m]bolic messages for the cameras, and ultimately for the

rest of America. Everyone should make their best effort to dress up . . . . Yes, thats right . . . Dress up as patriotic or famous figures such as Uncle Sam, or police officers, or real patriot politicians with white whigs, or judges . . . Plenty of Reagan face masks! **Deface masks the best you can**. This, and SUPRISE is our best element. If the piggies know where our rally is going to be at, they are going to meet us there, with hand-cuffs, rubber bullets, and tear gas. (BTW. It would also be great for us to have gasmasks of our own! But with BOWLS ATTACHED! Yup, gas mask bongs. Mimic the piggies.) I was also thinking that some insane cops could go as far as trying to incarcerate us and prosecute us as (the dreaded T-word) TERRORISTS! That would get a lot of publicity. You have to understand that the public mostly supports the government from what they see through the boobtube. Whatever the government tells them, is what is important. The government (minus the FCC) can not control the media under the right of Freedom of the Press. Anyway. This is a great great idea, and this is when the American government really does not want to talk about a not-so-important topic such as the Drug War. Kind of like when your parents were busy when you were little, and you kept bothering them, and bothering them, and bothering them. They don't appreciate that. They get pissed (and often gave in). We've finally got a lot of motivation. DO NOT LET IT DIE OUT! Make fliers, posters, make phone calls, send emails, send mail, knock on doors, and most importantly . . . . Talk to dealers and make sure they can actually supply all of us with some weed! We can't have a smoke out unless we have weed! BTW. To prevent ultra-prosecution and basically personal money loss, only bring 2-3 joints to the organized smokeouts. The cops are going to take them. Why waste your money on a half-o when the cops are going to take it anyway (and smoke it!)?
Stay motivated,
Fight for MMJ and legality,

NITSUJ
2/8/03
I don't smoke anymore. I am for the cause. And i am in high school. I will be 18 next presidential election so Bush better choose his words wisely. support the green party. thanks for reading all that

Message #234 posted by Holden99 September 25, 2002 22:55:33 ET
I'd love to organzie a get together of supporters in Iowa or surrounding states to voice and live what we believe—My vision is something very small at first, just a gathering of a few trusted, known and politically active people to spend some time to brainstorm, empower and maybe get a little high if that's your thing. Eventually, if it succeeds we can slowly make it bigger and bigger and slowly adapt and influence the mainstream. America isn't ready yet but grassroots with a vision can move mountains. Email if you're interested or have ideas.

Message #115 posted by SweetLiberty2 (Info) January 15, 2002 23:56:32 ET
As part of a Nationwide Bill of Rights Campaign, a rally focusing on how our freedoms have been eroded over the past 2 centuries by politicians and agencies acting outside the law, will be held this Saturday (Jan. 19) in Denver. Starting at High Noon we will be gathering in Lincoln Park which is across the street from the West steps of the state Capitol. Topics will include the new "Patriot

Act," the illegal War on Drugs, unlawful searches and arrests by Law Enforcement Officials, and more . . . . Come stand up for your rights. A large crowd
will generate media attention which could finally get the truth out to the public
about the state of our nation. Come hear what your government is REALLY up
to, what the media won't tell you, and what you can do about it. See you there!
Denver Bill of Rights Rally Saturday, January 19, 2002—High Noon in Lincoln Park across the street from the West Steps of the state Capitol.

These brief excerpts from the coded sites show us the range of how folks discuss
drugs in relation to music—or more accurately, how they don't. The majority of
these websites host a variety of cultural information and interaction, some of
which overlaps with either music or drug use, or both. However, assessing how
much these two parts of a varied cultural spectrum are linked is quite difficult.
Rave and its attendant scene is not the only music, and drugs are not the only
recreational or mind-altering substances/practices that those who submit links to
the Reciprocal Links of Erowid focus upon. Erowid is a trusted source for drug
information—both positive and negative—16% (135/854) of our survey respondents mentioned Erowid in answering our question, "A lot of people put information about drugs out on the web. How do you decide which sites offer the
most accurate and reliable information? What are the characteristics of a good
drug information/education website?" But my perusal of Erowid (see Table 7.1
above) shows the overlapping of that major topic with others, as well as a complex mix of discussion, action, and contemplation.

Discussion: Linked, but Connected? Connected, but Influ-
enced? Influenced, but Acting?

From the perspective of being immersed in a scene that encompasses online
interaction with information and the people who make and disseminate it, it is
crucial to understand the intensity, purpose, and effectiveness of such communications. In order to fully engage with the ethnographic method in this research
project, it was necessary to begin at the beginning—at least from one point of
entry on this scene, the Internet. There is an articulated concern with a particular
population—youth—and their uses of the 'net for reasons that range from worries over media over-immersion, sexual predators, and initiation into drug usage.
As I hope I have shown herein, I think it is only by combining an assessment
method that attempts to take into account the shape and density of the network
(the Internet, or a particular subset of it) with a participant observation method
that both reads the 'net, and attempts to contact those who make up the 'net—
folks like The Gus and others—that we can get at the typical and normal practices of the "scene."

The information and interactive website network in which the
rave/dance/drug-using scene is at least partly embedded is also nested within a
larger and ongoing debate about use and abuse, and the legalization of particular

substances, and whether exposure to deviant cultures—or even to their images—cause deviant behavior.[5] As well, this discourse is linked to the ongoing debates about other legal issues—such as free speech, privacy, and democracy—on the Internet and in society generally (See Demac and Downing, 1995; Demac and Sung, 1995; Etzioni, 1992, 1998a, 1998b; Gandy, 1995; Godwin, 1998; Goldring, 1997; Gurak, 1997; Hickok, 1991; Lessig, 1999; Lipshulz, 2000; Loader and Thomas, 2000; Mackinnon, 1998; Mayer-Schonberger and Foster, 1997; Reidenberg; 1997; Rheingold, 1993; Riley, 1996; Schiller, 1995; Sclove, 1995; Smith, McLaughlin, and Osbourne, 1998; Sobel, 1999; Sutton, 1996; Walzer, 1983, 1992, 1995; Wallace and Mangan, 1997).

In order to fully flesh out my tentative conclusions, a matched method assessment and coding of other clearinghouse websites such as DanceSafe and Students for Sensible Drug Policy would be the best route. We could then take those assessments and compare them with self-reported online contact and drug use links, and with the field research presented herein. Thus, we could assess ease of navigation and links from online to offline contact in several ways, from several points of entry. Our survey included several open-ended questions along with several interval-level questions designed to gauge both density and content of drug-related Internet-based networks:

25c. In your opinion, how important are drugs and alcohol to the "rave" community? At the parties/raves you attend, what percentage of people there do you think are drunk or high?

25d. From what you've seen at raves/dances and clubs, how do you think music affects someone's drug experience? **(Please give a specific example.)**

25e. What types of music seem to go best with specific types of drug experiences?

45. A lot of people put information about drugs out on the web. How do you decide which sites offer the most accurate and reliable information? What are the characteristics of a good drug information/education website?

46. Why do you participate in discussion boards or chat rooms? What do you enjoy talking about the most?

47. If you could sit down with us and tell us anything that you think we should know about recreational drug use, music and raves, online communities, or the government's policies toward drugs, what would you say?

52. Are there any drugs that you particularly enjoy taking together? If so, please list them and tell why you like this combination. Where did you learn about this combination? (Murguía, et al., 2003).

Figuring out these issues of shape, density, content, and contact for people who actually engage with a complex scene is best served by an ethnographic method understood as an extended case method, one that is attuned to the inherently multi-sited enterprise of actions and interactions undertaken in Internet settings (see Burawoy, 1991; 2000; Marcus, 2000; Gatson and Zweerink, 2004a; 2004b:29-30). We have to follow the information to its source(s), and follow the people across online and offline terrains to best capture their practices.

Notes

1. As Farrell and Carpini discuss (2004), there are emergent ways to map Internet interactions that pay particular attention to the important nodes in an interactive network, stressing the flow of information between particular sites, although this method is not able to map according to content, nor in terms of actual use that people make of the links, either online or offline.

2. See the Association of Internet Researchers' "Ethical decision-making and Internet research" for a discussion on how various Internet venues establish their own ethical expectations (2002: 4-5).

3. I did not count sites in which only email contacts were offered as significant CMC, although this may indeed be a gateway. To date, I have only looked at sites with significant traffic available for public consumption. In addition, the sites coded here are those sites that were actually accessible and active on the days that I was coding sites. There is some turnover of sites, and there are often hardware issues at either my end of the cyberportal, or that of the sites themselves. Sites that are in fact in working order on other days may thus be missed herein.

4. In Gatson, 2004a: 163, I highlighted a subset of the fifty-two sites—twenty-one sites that were consistently and always active. Of these sites, only 38% were coded on the +1-+3 end of the scale for the "Contact → likelihood of drug use" category. Concern regarding any particular site's effect on behavior thus fluctuates with that site's availability.

5. These discussions range from worries over apathetic, "dumbing-down," and retreatist forms of deviance to concern over the aggressive actions caused by mass media and mass culture (see Horkheimer and Adorno, 1972; Mander, 1978; Neuhaus, n.d.).

References

Burawoy, Michael. *Ethnography Unbound: Power and Resistance in the Modern Metropolis.* Berkeley, CA: University of California Press, 1991.
———. *Global Ethnography: Forces, Connections, and Imaginations in a Postmodern World.* Berkeley, CA: University of California Press, 2000.
Berger, Harris M. "Discussion of networks, scenes, and communities." Personal communication, c2001.
Conrad, Barnaby. *Absinthe: History in a Bottle.* San Francisco: Chronicle Books, 1997.
Crispino, Azzurra. "Deterrence of Harm to Self: A Study of Online Rhetoric." in *Real Drugs in a Virtual World: Drug Discourse and Community Online,* edited by Edward Murguía, Melissa Tackett-Gibson, and Ann Lessem. Lanham, Md.: Lexington Books, 2007.

Demac, Donna, and John Downing. "The Tug-of-War Over the First Amendment." Pp. 112-27 in *Questioning the Media: A Critical Introduction.*" edited by John Downing, Ali Mohammadi, and Annabelle Sreberny-Mohammadi. Thousand Oaks, CA, London and New Delhi: Sage Publications, 1995.

Demac, Donna A., with Liching Sung. "New Communication Technologies and Deregulation." Pp. 277-92 in *Questioning the Media: A Critical Introduction.*" edited by John Downing, Ali Mohammadi, and Annabelle Sreberny-Mohammadi. Thousand Oaks, CA, London and New Delhi: Sage Publications, 1995.

Durkheim, Emile. *Suicide.* New York: The Free Press, 1951 (1979).

————. "Ethical decision-making and Internet research." Association of Internet Researchers Ethics Working Committee, 2002.

Etzioni, Amitai. *The Spirit of Community: Rights, Responsibilities, and the Communitarian Agenda.* New York: Touchstone/Simon and Schuster, 1992.

————. "Old Chestnuts and New Spurs." Pp. 16-36 in *New Communitarian Thinking: Persons, Virtues, Institutions, and Communities.* edited by Amitai Etzioni. Charlottesville and London: University Press of Virginia, 1995.

————. "A Matter of Balance, Rights and Responsibilities." Pp. xi-xxiv in *The Essential Communitarian Reader.* edited by Amitai Etzioni. Lanham, MD: Rowman and Littlefield Publishers, Inc, 1998.

————. "The Responsive Communitarian Platform: Rights and Responsibilities." Pp. xxv-xxxix in *The Essential Communitarian Reader.* edited by Amitai Etzioni. Lanham, MD: Rowman and Littlefield Publishers, Inc, 1998.

Farrall, Kenneth N., and Michael X. Delli Carpini. "Cyberspace, the Web Graph and Political Deliberation on the Internet." *PISTA 2004 Proceedings*, Vol. I. Edited by Jose V. Carrasquero, Friedrich Welsch, Angel Oropeza, Charles Mitchell, and Maritta Välimäki. Orlando, FL: International Institute of Information and Systematics (2004): 287-94.

Gandy, jr., Oscar H. "Tracking the Audience: Personal Information and Privacy." Pp. 221-37 in *Questioning the Media: A Critical Introduction.*" edited by John Downing, Ali Mohammadi, and Annabelle Sreberny-Mohammadi. Thousand Oaks, CA, London and New Delhi: Sage Publications, 1995.

Gatson, Sarah N. "Illegal Behavior and Legal Speech: Discourse about Drug Use in an Internet Community/Network." Paper presented at Southwestern Social Science Association Meetings, San Antonio, TX, 2003.

————. "Youth, Drugs, Internet Discourse: A new arena of community and citizenship." *PISTA 2004 Proceedings*, Vol. I edited by Jose V. Carrasquero, Friedrich Welsch, Angel Oropeza, Charles Mitchell, and Maritta Välimäki. Orlando: International Institute of Information and Systematics (2004): 163-68.

Gatson, Sarah N., and Amanda Zweerink. "Choosing Community: Rejecting Anonymity in Cyberspace." *Research in Community Sociology*, vol. 10 (2000): 105-37.

————. "'Natives' Practicing and Inscribing Community: Ethnography Online." *Qualitative Research*, 4, no. 2 (2004a): 179-200.

————.*Interpersonal Culture on the Internet—Television, the Internet, and the Making of a Community.* Lewiston, NY: The Edwin Mellen Press: Studies in Sociology Series, no. 40. 2004b.

Godwin, Mike. *Cyber Rights: Defending Free Speech in the Digital Age.* Cambridge, MA: Massachusetts Institute of Technology, 1998 (2003).

Goldring, John. "Netting the Cybershark: Consumer Protection, Cyberspace, the Nation-State, and Democracy." Pp. 322-54 in *Borders in Cyberspace: Information Policy and the Global Information Infrastructure*. Cambridge, MA and London: MIT Press, 1997.

Gurak, Laura J. *Persuasion and Privacy in Cyberspace: The Online Protests over Lotus Marketplace and the Clipper Chip*. New Haven, CT: Yale University Press, 1997.

Hickok, jr., Eugene W., editor. *The Bill of Rights: Original Meaning and Current Understanding*. Charlottesville, VA and London: University of Virginia Press, 1991.

Hirschi, Travis. *Causes of Delinquency*. Berkeley, CA: University of California Press, 1969.

Horkheimer, Max, and Theodore Adorno. "The Culture Industry: Enlightenment as Mass Deception." in *The Dialectic of Enlightenment*. New York: Harder and Harder, 1972.

Kotarba, Joseph A. 2003. "Music as a Feature of the Online Discussion of Illegal Club Drugs." Typescript.

Lessem, Ann. 2003. "The Neverending Conversation: A Case Study of Rave-Related Internet Conversation and Drug Use." Typescript.

Lessig, Lawrence. *Code, and Other Laws of Cyberspace*. New York: Basic Books, 1999.

Lipschultz, Jeremy Harris. *Free Expression in the Age of the Internet: Social and Legal Boundaries*. Boulder, CO: Westview Press, 2000.

Loader, Brian, and Douglas Thomas, editors. *Cybercrime: Law Enforcement, Security and Surveillance in the Information Age*. New York: Routledge, 2000.

MacKinnon, Richard. "The Social Construction of Rape in Virtual Reality." Pp. 147-72 in *Network and Netplay: Virtual Groups on the Internet*. edited by Fay Sudweeks, Margaret McLaughlin, and Sheizaf Rafeli. Menlo Park, CA, Cambridge, MA, London, UK: AAAI Press/MIT Press, 1998.

Malley, Wendi. 2003. "Blogging—A Cyber Portal: Changing Audiences to Publics." Typescript.

Mander, Jerry. *Four Arguments for the Elimination of Television*. New York: Quill, 1978 (1977).

Marcus, George. *Ethnography Through Thick and Thin*. Princeton: Princeton University Press, 2000.

Mayer-Schonberger, Viktor and Teree E. Foster. "A Regulatory Web: Free Speech and the Global Information Infrastructure." Pp. 235-54 in *Borders in Cyberspace: Information Policy and the Global Information Infrastructure*. Cambridge, MA and London: MIT Press, 1997.

Merton, Robert K. "Social Structure and Anomie." *American Sociological Review*, Vol. 3, no. 5 (1938.): 672-82.

Mitchell, William J. *E-Topia: Urban life Jim, but not as we know it*. Cambridge, MA and London, UK: The MIT Press, 2000.

Murguía, Ed, Sarah N. Gatson, and Melissa Tackett-Gibson. "Youth, Technology, and the Proliferation of Drug Use." Grant, National Institute on Drug Abuse/ National Institute of Health, 2001.

Murguía, Edward, Sarah N. Gatson, Melissa Gibson, Joseph Kotarba, Ann Lessem, Kamesha Spates, and Rachel Willard. "Youth, Technology, and the Proliferation of Drug Use." Final Report, National Institutes on Drug Abuse/National Institutes of Health. 2003.

Neuhaus, Richard John. "The Internet Produces a Global Village of Village Idiots."(http://www.fpc.cc.tx.us/~CLASSES/OFSO1313/index5.htm#Global%20Idiots) Retrieved November 15, 2003.

Pinchbeck, Daniel. *Breaking Open the Head: A Psychedelic Journey into the Heart of Contemporary Shamanism.* New York: Broadway Books, 2002.

Redden, Jim. *Snitch Culture: How Citizens are Turned into the Eyes and Ears of the State.* Venice, CA: Feral House, 2000.

Reidenberg, Joel. "Governing Networks and Rule-Making in Cyberspace." Pp. 84-105 in *Borders in Cyberspace: Information Policy and the Global Information Infrastructure.* Cambridge, MA and London: MIT Press, 1997.

Rheingold, Howard. *The Virtual Community: Homesteading on the Electronic Frontier.* Reading, MA: Addison-Wesley Publishing Company, 1993.

Riley, Donna M. "Sex, Fear and Condescension on Campus: Cybercensorship at Carnegie-Mellon." Pp. 158-68 in *Wired Women: Gender and New Realities in Cyberspace.*" Seattle, WA: Seal Press, 1996.

Schiller, Herbert I. "The Global Information Highway: Project for an Ungovernable World." Pp. 17-34 in *Resisting the Virtual Life: The Culture and Politics of Information.* edited by James Brook and Iain A. Boal. San Francisco: City Lights, 1995.

Sclove, Richard E. "Making Technology Democratic." Pp. 85-104 in *Resisting the Virtual Life: The Culture and Politics of Information.* edited by James Brook and Iain A. Boal. San Francisco: City Lights, 1995.

Smith, Christine B., Margaret L. McLaughlin, and Kerry K. Osbourne. "From Terminal Ineptitude to Virtual Sociopathy: How Conduct is Regulated on Usenet." Pp. 95-112 in *Network and Netplay: Virtual Groups on the Internet.* edited by Fay Sudweeks, Margaret McLaughlin, and Sheizaf Rafeli. Menlo Park, CA, Cambridge, MA, London: AAAI Press/MIT Press, 1998.

Sobel, David, editor. *Filters and Freedom: Free Speech Perspectives on Internet Content Controls.* Electronic Privacy Information Center, 1999.

Steinberg, Laurence, and Laura H. Carnell. "Testimony before the Subcommittee on Crime of the U.S. House Judiciary Committee." American Psychological Association Congressional Testimony. 1999. <http://www.apa.org/issues/psteinberg.html> (30 Apr. 2003).

Steinberg, Laurence. "Should Juvenile Offenders be Tried as Adults? A Developmental Perspective on Changing Legal Policies." January 19, 2000.<http://www.jcpr.org/authors_otherwork/steinbergpapers.html> (30 Apr. 2003).

Sutton, Laurel A. "Cocktails and Thumbtacks in the Old West: What Would Emily Post Say?" Pp. 169-187 in *Wired Women: Gender and New Realities in Cyberspace.*" Seattle, WA: Seal Press, 1996.

Talmadge, Joe. "The Flamers Bible." 1987. <http://www.netfunny.com/rhf/jokes/88q1/13785.8.html> (Aug. 1999).

The Vaults of Erowid. "Documenting the Complex Relationship Between Humans and Psychoactives." <http://www.erowid.org/> (2002-2003).

Tönnies, Ferdinand. *Community and Society (Gemeinschaft und Gesellschaft).* New York: Harper and Row. 1963 (1887).

Walzer, Michael. 1983. *Spheres of Justice: A Defense of Pluralism and Equality.* New York: Basic Books.

———. *What it Means to be an American.* New York: Marsilio Publishers Corp, 1992.

———. "The Communitarian Critique of Liberalism." Pp. 53-70 in *New Communitarian Thinking.* edited by Amitai Etzioni. Charlottesville and London: University Press of Virginia, 1995.

Wallace, Jonothan, and Mark Mangan. *Sex, Laws, and Cyberspace: Freedom and Censorship on the Frontiers of the Online Revolution.* New York: Henry Holt, 1997.

Zimmer, Lynn and John P. Morgan. *Marijuana Myths Marijuana Facts: A Review of the Scientific Evidence*. New York and San Francisco: The Lindesmith Center, 1997.

Zweerink, Amanda, and Sarah N. Gatson. "WWW.Buffy.Com: Cliques, Boundaries, and Hierarchies in an Internet Community." Pp. 239-49 in *Fighting the Forces: What's at Stake in Buffy the Vampire Slayer*. edited by Rhonda Wilcox and David Lavery. Lanham, Md: Rowman and Littlefield, 2002.

Chapter Eight

Scripters and Freaks: Knowledge and Use of Prescription Stimulants Online

Melissa Tackett-Gibson

Prescription drug abuse has gained attention in recent years due to notable increases in the use of pain killers and barbituates. Accounts in the popular press of illicit Oxycodone and Hydrocodone use have highlighted the problem. According to the National Survey on Drug Use and Health, nearly twenty-one million Americans over the age of twelve have taken prescription medications, such as pain relievers, sedatives, and stimulants for nonmedical purposes (SAMHSA, 2005a; SAMHSA 2005b). Over six million have used pharmaceuticals in the past year for recreational purposes; and particularly high rates of use are found among eighteen through twenty-five-year-olds. Within this age group the prevalence of prescription drug abuse was three times higher than that of cocaine, four times higher than hallucinogens, and twice that of ecstasy. A recent SAMHSA report similarly found that across all age groups, psychotherapeutics were abused at a higher rate than cocaine, hallucinogens, and inhalants (SAMHSA, 2003). Unfortunately, the levels of use have been consistent over time. This has led some to conclude that current levels of nonmedical use show "no signs of decreasing among young adults," (SAMHSA, 2005a).

While outpaced by the abuse of prescription pain relievers and tranquilizers, stimulant abuse is also currently high. This is typically attributed to increases in the availability of drugs used to treat attention deficit disorder. According to the national surveys, approximately 800,000 persons are introduced to the abuse of stimulants per year; and 1.2 million currently use stimulants (SAMHSA, 2005c). Since 2002 approximately 10% of the population has reported nonmedical use of stimulants (SAMHSA, 2003). Specifically, the nonmedical use of Ritalin increased from 0.1% in 1992 to 2.2% in 2000. This rate doubled by 2005 to 4.4%.

Like other prescription drugs there are advantages to using and procuring prescription stimulants. The production and distribution of the substances are controlled—psychotherapeutics are consistent in dosage and purity; the acquisition of stimulants can be "legal" with a prescription; and the costs of the drugs are often subsidized by insurance policies. Online discussion board transcripts suggest that young adults engaged in drug-related talk routinely share information about prescription medications. In particular, they discuss the methods of obtaining prescriptions (i.e., "scripting," "doctor shopping"); and the recreational value of medications. Drugs such as Ritalin, Adderall, and Dexedrine are the focus of many such discussions.

This paper qualitatively examines stimulant use and abuse among participants in an online drug information forum. It highlights the themes of discourse that consistently appeared over time, and explores perceptions of prescription stimulant use and users. It focuses on transcripts from forum discussions of prescription stimulants such as methylphenidate (Ritalin, Concerta), amphetamine (Adderall), dextroamphetamine (Dexedrine), and methamphetamine (Desoxyn). These drugs are considered central nervous system stimulants and have a high potential for abuse. While this article is primarily descriptive it provides qualitative data that compliments recent quantitative research on stimulant abuse. Most importantly, it provides insight into issues of access and the value of prescription stimulants from the user's perspective.

Stimulant Use as a Drug Abuse Issue

Recreational stimulant use is not new. In 1930 amphetamines were first marketed for medical use under the brand name Benzadrin and by the 1940s they were commonly used in the treatment of congestion and narcolepsy (DEA, 2003). As medicinal use of amphetamines became more common so did the misuse and abuse of the drugs for recreational purposes. The rates of abuse peaked during the 1960s and 1970s as the illegal production and medical use of amphetamine and methamphetamine became widespread. Government controls and revised prescription practices curbed amphetamine misuse until recent years. However, with new interest in the treatment and diagnoses of ADD/ADHD, the use of prescription amphetamines is again common medical practice.

Generally, the use of stimulants under a physician's supervision poses few health risks. However, the nonmedical use and abuse of stimulants is potentially dangerous. Like illicit amphetamines and methamphetamines, prescription stimulants can cause irritability, paranoia, hallucinations, seizures, hypertension, and tachycardia when used in excessive amounts (Klein-Schwartz, 2002; NIDA,

2001; DEA, 2003). Addiction to stimulants is also of risk in the case of excessive and frequent use. Therefore users may also experience significant withdrawal symptoms upon the cessation of use. When used with other illicit drugs, some anti-depressants, and antihistamines compound the adverse effects of misused prescription stimulants (NIDA, 2001).

Today, stimulants are commonly prescribed to treat an array of medical conditions in addition to ADD/ADHD. These include asthma, narcolepsy, obesity, and depression. Apart from their legitimate medical uses, legal stimulants have considerable abuse potential due to drug effects such as increased perceived alertness and euphoria. Researchers and medical practitioners have long been concerned about the abuse of and addiction to prescription stimulants (Klein-Schwartz and McGrath, 2003). Given recent increases in the diagnoses of ADD/ADHD among adolescents, there is renewed speculation that prescription stimulants are commonly diverted for nonmedical use. Accordingly, the topic of prescription stimulant abuse has received a good deal of attention in recent research.

Current studies have examined the propensity for abuse among those presenting ADD/ADHD symptoms and diagnoses (Clure, et al., 1999; Levin and Kleber, 1994; Coetzee, Kaminer, and Morales, 2002); alternative treatment models and medications for individuals likely to abuse (Stitzer and Walsh, 1997); and the likelihood of diversion of medications for nonmedical use (Musser et al., 1998). Several notable studies have specifically examined the abuse of prescription stimulants among adolescents and young adults (Poulin, 2001; Musser, et al. 1998; Babcock and Byrne, 2000; Graff Low and Gendaszek, 2002). On the whole, they suggest that prescription stimulant abuse is cause for concern and further research. They suggest that diversion is common, that the rates of abuse are increasing over time, and that environmental conditions specific to adolescences and young adulthood may contribute to abuse.

Poulin examined the use of stimulants among high school students, with special attention to the diversion of medically prescribed stimulants for recreational use. Her work found that a higher proportion of students had used prescription stimulants for recreational purposes than for legitimate medical treatment (8.5% compared to 5.3% of the students surveyed). Diversion of stimulants was common—more than 20% of the students who had been prescribed stimulants had given or sold them to other students at some time. Results also indicated that students were more likely to engage in the nonmedical use of stimulants if they were in classes where students were willing to give or sell their medically prescribed stimulants, suggesting that classroom relationships serve to facilitate the diversion of stimulants. Use of other substances, such as alcohol, cigarettes, and marijuana, was also found to be a risk factor associated with nonmedical stimulant use.

Similarly, in their study of reports of harmful substance exposures, Klein-Schwartz and McGrath (2003) concluded that the abuse of prescription stimulants, specifically methylphenidate, was on the rise. Using exposure data col-

lected from the American Association of Poison Control Center's Toxic Exposure Surveillance System (AAPCC-TESS), researchers tracked the incidence of
exposure cases from 1993 to 1999. They included only those cases of exposure
that resulted from intentional abuse, as well as those that involved the abuse of
methylphenidate and additional substances. Among youth ages ten through nineteen, the study found that 11% of exposure reports resulted from the abuse of
methylphenidate. Of those a substantial majority (70%) of the reports involved
the sole abuse of methylphenidate with the remainder resulting from the abuse
of polysubstances. The number of abuse-related reports for that age group increased over time with only 17 cases reported in 1993 compared 158 in 1998.
Among all ages methylphenidate abuse report increased ninefold within the
same time period.

The authors state, the study is particularly relevant in that it provides further evidence that not only is prescription stimulant abuse a growing problem,
but that methylphenidate abuse is not limited to pre-teens and adolescents. They
also note, however, that the study has shortcomings. While overdose reports to
poison control centers are solid indicators of abuse trends, not all cases are reported. Contact with the centers is voluntary and may be likely to occur among
those with little previous exposure to drug effects. Therefore, the abuse habits of
chronic users are likely overlooked. Similarly, study data does not shed light on
factors leading to use, frequency of use, or the dosages taken prior to the report.
While their analysis is critical to the study of prescription stimulant abuse it is
only a part of the full picture of abuse.

Unlike the two previous studies, the following highlight abuse specifically
among university and college students. In their study of stimulant abuse, Graff
Low and Gendaszek (2002) surveyed 150 young adults at a competitive private
college. More than a third (35%) of those surveyed reported the use of prescription stimulants, such as methylphenidate, without a prescription. 10% reported
monthly use and 8% indicated that they used weekly. When asked in open-
ended question the reasons for use, students cited performance enhancement in
work, exercise, and athletics. According to the authors, these motivations are
particularly strong among young adults in a college setting. Strong motivations
to achieve and seek new sensations, along with the ease of obtaining prescription stimulants, create an atmosphere where use is common and problematic
among students.

The study is particularly relevant to this research in that it specifically examines young adults. As Graff Low and Gendaszek note, however, it is limited
in that it defined those with legal prescriptions as legitimate users. It did not
examine the misuse and abuse patterns of those diagnosed with conditions
treated with legal stimulants. Moreover, it assumed that respondents were diagnosed as a result of the actual presence of a condition and not drug-seeking behaviors. This ignores a critical characteristic of prescription drug abuse—the
procurement of illegal substances through legal avenues.

In a similar study Babcock and Byrne (2000) surveyed nearly one thousand
five hundred students at a liberal arts college. Of those that responded (20%),

over half reported knowing someone that had used Ritalin for recreational purposes. While slightly lower, the rates of Ritalin use were comparable to those of illicit stimulants such as cocaine and amphetamine (16% compared to 22% and 24% respectively). Traditional students (those twenty-four and younger) were much more likely than their older counterparts to use Ritalin for nonmedical purposes (20.9% versus 3%). Few respondents had a prescription for Ritalin and a third (45% of younger students) knew someone from whom they could purchase the drug. Intranasal administration of the drug was commonly reported.

The authors suggest that use may be affected by environmental circumstances particular to a college setting. They cite the pressures of exams and frequent deadlines as reasons students may turn to the use of stimulants. The data, however, provide no indication that this is the case. In fact the survey was limited to ten questions and provides little detailed information related to use patterns. Two important aspects of use, frequency of use and dosage, were omitted. Respondents may be reporting high rates of one-time use or instances of use in small amounts. Therefore the findings may overestimate the levels of use across the campus.

Again, all of the cited studies provide valuable data related to the abuse of psychostimulants. However, each offers a look at the problem of abuse specifically through the lens of quantitative methodologies. Poulin, Babcock and Byrne, and Graff Low and Gendaszek used survey techniques. Klein-Schwartz and McGrath relied upon available data. Many of the details of use and abuse patterns were necessarily unaddressed. A qualitative study of the issue can uniquely contribute to the body of knowledge on the subject in that it can richly characterize use, factors related to the causes of use, and methods of use specifically from the user's perspective.

Research Design and Methods

While the nonmedical use of psychotherapeutic medications has increased in recent years, so has the use of the Internet for health and drug information. As discussed in my previous chapter, online communication can quickly inform one about prescription drugs, their health impacts and risks, and their recreational effects. Currently, there are approximately ten million active websites with content devoted to Ritalin. They range from the federal government's own drug information archives to online pharmacies and ADHD support groups. In addition, prescription stimulants are easily found referenced on sites with drug use content. In the past year, "Ritalin" has been the topic of over six thousand discussion threads on the website that is the focus of this study, DrugSite. A similar harm reduction website hosted 3,100 threads on the topic. Erowid devotes several information pages with a plethora of substantive links related to the drug and other amphetamines such as dextro-amphetamine. Information about prescription stimulants is ubiquitous online.

Research data for this study were collected from January 2002 to April 2003. Data were obtained from transcripts of online threaded discussions that took place at a large popular drug information website, "DrugSite." Two forums provided the text for this study: "Pill Primer" and "The Harder Stuff." Prescription stimulants were discussed in other forums, such as those dedicated to health and well-being, and reflection on the negative effects of use. However the overwhelming majority of threads related to stimulant use appeared in the drug use specific forums. It should be noted again that in the context of all of the discussion on the site the ten drug use forums accounted for only approximately 10% of the sites discussion activity. Thus while discussions of stimulants were predominately found in drug use forums, they were not widespread on the site.

The forums, "Pill Primer" and "The Harder Stuff" mainly included discussions of drugs such as Ecstasy and other illicit drugs, including prescription medications. The first board, "Pill Primer" was primarily comprised of those participants rather new to drug use. They discussed traditional club drugs and typically discussed pharmaceuticals in light of club drug use. For example, forum members typically expressed interest in the reactions between prescription stimulants and other illicit drugs. Those involved in "The Harder Stuff" were typically more experienced users. In their discussions of stimulants they actively discussed methods of obtaining the drugs, the amount of drug necessary for effect, and other topics specifically related to the nonmedical use of stimulants.

During the research period many "threads" or forum discussions were dedicated to prescription drugs in general and stimulants in particular. This resulted in copious amounts of data for analysis. To narrow the scope, transcripts of the discussions were analyzed only if they were substantive and "popular"—if a large number of site members participated. Popular discussions were defined as those that solicited input from approximately eight to ten members. Ultimately fifty threads, specifically dealing with prescription drug use topics, were collected. Eight addressed issues related to the use of prescription stimulants and were selected for in-depth analysis. Analyses centered on the discourse surrounding prescription stimulants—the characterizations of prescription stimulant use and users; and the narratives common to use, such as those about obtaining prescriptions and the benefits of use.

Both males and females participated in the selected discussions, although due to the anonymous nature of online communication it was difficult to determine the number of men or women that contributed to the discourse. Site members including "newbies," regular participants, moderators, and administrators participated in the threaded discussions. In all 115 individuals contributed to the selected discussions, of those 13 posted in more than one of the threads. Nearly all of those that posted directly referenced personal recreational prescription drug use or the intention to use in the future. Several also reported a diagnosis of attention deficit disorder in the past, however the extent to which they suffered from symptoms was difficult to determine from the texts.

Major Themes of Discourse

Methods of and Reasons for Nonmedical Stimulant Use

Qualitative evidence from discourse transcripts confirms many of the findings of previous research. Similar to respondents in the Babcock and Byrne study (2000) discussion participants cited crushing and snorting Ritalin and Adderall. In some cases, "parachuting" the drug was recommended. This was done by folding crushed Adderall pellets into toilet paper and then ingesting the "parachute." This method allowed the drug to be quickly absorbed and the affects quickly felt. Though not nearly as popular and often considered more dangerous, intravenous use was also cited as a means of use.

Similar to the other studies participants cited increased concentration for studying, going to class, or work as a reason for use. For example, one forum participant consistently used Ritalin to improve his performance in class and increase his productivity. Another considered stimulant use wiser than the use of marijuana or opiates due to the fact he could work while using the stimulant.

> At least when you're tweakin you get the job done. Try getting a drunkard, stoner or heroin addict to clean the reverse side of a ceiling fan.
> (Tweaky_Bird, April 2003)

Several also preferred prescription stimulants to "street" drugs because of access routes and drug purity. They cited the inability to locate methamphetamine or other illicitly produced stimulants, the low cost of acquiring the drugs, and the consistency of drug content as important factors in the decision to use pharmaceuticals. In a debate over the virtues of Adderall, a participant conceded that methamphetamines would probably provide a better "rush." However, his access to illicit stimulants was limited.

> I know Adderall is nothing compared to meth but I cant find meth.
> (Rainman, April 2003)

In the same thread Rainman also argued that Adderall was a wiser choice due to its purity. Other participants agreed with his assessment and one even encouraged users to choose prescription stimulants over street drugs.

> If you gotta have speed . . . good move stickin with the adderall . . . don't go lookin for meth. youll be a goner in six months. I don't know if you realize the garbage that goes into that stuff and the dudes cooking it aren't the sharpest tools in the shed.
> (Squidward, April 2003)

Few found costs prohibitive to their use of the drugs. In fact some reported that insurance coverage subsidized a major proportion of the cost. One participant ambivalently argued that access to the drug was too easy. He described his prescription and the associated costs of acquiring the drug.

> 360 5mg dextroamphetamine pills per month (cost $4 co-pay x 12)
> (Prankster999, April 2003)

Access to Prescription Stimulants

It was the ease of access to the drugs that fueled much of the conversation. Participants cited various methods of obtaining the substances, including theft and diversion. For example, one participant reported stealing the drug from the school nurse's office. Others reported repeated incidents of drug diversion. They reported that they had received drugs from friends or acquaintances with stimulant prescriptions. AsianKing's report of getting drugs from his friend was a typical account of stimulant diversion:

> [a friend] gave me her bottle of about 25 10mgs, which . . . are about gone.
> (AsianKing, January 2002)

While many stimulant users reported access through friends, ultimately the respondents considered an ADHD diagnosis to be the easiest and quickest means of obtaining drugs. Several of the most popular narratives involved discussions related to finding a doctor and obtaining a prescription for a stimulant. In these narratives participants were coached to "doctor shop"—meaning that they were instructed to find a doctor who liberally writes prescriptions. They were also encouraged to develop a plausible story that would convince a doctor of the need for the medication. Thus participants were often coached in art of gaining a prescription, or in "scripting."

For example, Diskojunky asked board members for advice on getting a prescription for stimulants:

> Id (sic) like to get a prescription for adderall or Ritalin. Ive been buying it from a friend for cheap, but ill save money if I get prescribed to it. what exactly should I say to my doctor.
> (Diskojunky, July 2002)

He was quickly reassured that he would eventually get a prescription. According to one participant, most doctors were liberal with Ritalin and Adderall prescriptions and finding one willing to prescribe the medication would not be difficult.

> They give scripts out for Adderall like they do AOL free trial membership CDs, . . . But if you think your current doctor wont consider anything but Ritalin why not visit another doctor. . . . You are almost guaranteed a script for Addies.
> (ACadaBra, July 2002)

There were also several methods cited for convincing doctors of the need for a prescription. Potential "scripters" were encouraged to learn about the symptoms of ADHD in order to clearly communicate their feigned problems to a doctor. They were also instructed to develop a personal history of the disorder and to act as if ADHD had been problematic since childhood. Others suggested informing doctors of a history of ADHD diagnosis and prescriptions stimulant use. StimBunny, AcadaBra, and Noestic commonly advised those who wanted a prescription:

> If you really want to fake ADHD go get a book on "coping" with the disorder. Read it through to get a general idea of what ADD is like . . . with a little acting ability you can convince most doctors you need powerful stimulant medication to make so you're able to focus at work or school. Hell bring the book in to your docs if that will help. You want to portray the guy who has a serious, genuine, not-faking-it-for-scripts medical condition . . . [you] want to get better!
> (StimBunny, July 2002)

> Imagine a pseudo-childhood for yourself and score some emotional points with whoever you see . . . a good actor shouldn't have problems getting as ADD script.
> (ACadaBra, January 2002)

> You gotta play this up, get in the mindset of "ok, I was diagnosed with ADHD and was taking medication for the last 6 years, now that I am no longer on it, I can't function" . . . that's how a good scripter pulls it off—once you have convinced yourself its all real, then you'll have no problems bullshitting a doctor.
> (NoesticO, January 2002)

Some participants reported putting the suggestions into practice and successfully getting prescriptions. One participant indicated that in order to get his prescription, he found a new doctor and falsely informed him of a previous diagnoses and drug protocol.

> Made an appointment with a new psych. I explained to him that my doctor in the previous city ordered every test in the book . . . my chronic fatigue was a complete mystery. Btw, like most doctors he didn't even write down the name of my previous doctor . . . who btw didn't exist. Told him my doc had prescribed every antidepressant in the book and nothing worked, then !Bingo! he tried Ritalin. 20mg of Ritalin twice a day changed my life. . . . Within seconds he breaks out the script pad. Success! . . . its all about finding the right doctor.
> (Filch, July 2002)

Another indicated that he had successfully received a new diagnosis and prescription, but he was now interested in getting a prescription for a stronger stimulant. The discussions indicated clearly that there was a hierarchy of stimulants of choice. The hierarchy was based on the type and formulation of various drugs, and it was perpetuated though online reports of the drugs' recreational effects. For example, Concerta, an extended release methylphenidate derivative, is very similar to Ritalin. However, forum participants believed the extended release coating left the drug with little recreational value. Thus Ritalin and Adderall, a mixture of amphetamines, were considered superior to Concerta.

> Score Concerta
> Im looking forward to getting my script bumped up to Adderall and I was wondering exactly what I should say.
>   (Corrupted, September 2002)

After obtaining the prescription, Corrupted was coached on how to get a prescription changed. Again the tactics were similar to those used to obtain a diagnosis. For example, Corrupted was advised to learn more about the side effects of Concerta. He was then encouraged to cite some of these side effects at his next doctor's visit. The timing of his request for stimulants such as Adderall or dexadrine was considered critical.

> [Next time you see the psychiatrist] you'll probably have your only chance to get Adderall or Dexedrine. You have to complain about some of the concerta specific side-effects . . . You should also mention the price . . . Adderall and dexadrine are cheap.
>   (ACadaBra, January 2002)

> AC is right, you need to make the move now . . . your doctor has shown a willingness to prescribe amphetamines, so this is a GOOD THING. If you enjoy Ritalin you will [like] Adderall or dexadrine more.
>   (NoesticO, January 2002)

Thus, gaining access to prescription stimulants was not difficult for those interested in use. Participants cited very few obstacles to finding stimulants through diversion or to getting prescriptions from doctors. In fact the most difficult part of the process described online was typically finding the type of stimulant desired.

Perceptions of Prescription Stimulant Users.

Lastly, in addition to topics of use and access, the perceptions of stimulant users were also discussed online. Prescription stimulant users were characterized as timid and weak by those who abused "street" methamphetamine and cocaine. Adderall and Ritalin users were characterized as "tiptoeing" around on the margins of drug abuse; as not having the courage to become "real" addicts. They were frequently described as children, "pussies," and pathetic. For example, ShukLuck and Raisn_DAZE goaded one participant for discussing his Adderall addiction:

Buck up kids! Go bang some speed. . . . [Adderall] is kinda like speed, only
without the balls. . . . if you're going to be an addict do it right
> (ShukLuck, April 2003)

As for all the Adderall addicts, step up you pussies! If you're going to abuse a
stimulant and f*** yourself then at least make it the king of stims. Adderalls
big ugly brother, Methamphetamine.
> (Raisn_DAZE, April 2003)

At times the abusive language toward prescription stimulant users was deleted
by board moderators. Participants scolded many of the hecklers and reminded
them to stop encouraging the abuse of harder substances. In general, prescription
stimulant users defended their use of pharmaceuticals and cited a preference for
the drugs.

Interestingly, it was within the context of the debate between prescription
and "street" stimulants that users frequently reflected on their use. They often
tried to distinguish between "recreational use" and addiction to prescription
stimulants. Some members were clearly considered addicts by other forum
members; others had ambivalent feelings about the levels of their own use. For
example one thread as a self-effacing joke began with the title, "you know you
are an Adderall addict when. . . . " While the thread was generally light hearted,
several users described in rather sad detail the consequences of their high levels
of use, such as the inability to sleep, loss of interest in eating, and an intense
psychological preoccupation with the drug. As they were confronted by other
concerned forum participants, several confronted their addictions. As one participant noted:

Thank God for it not being over the counter . . . the lack of availability is sometimes the only thing keeping me from being on it 24/7.
> (RADioFluke, April 2003)

Those with ADHD symptoms and diagnoses also questioned their motives for
taking the drug. This suggests that the role of the diagnosis made definitions of
recreational use ambiguous to some users. When asked if their use was simply
recreational or necessary for the control of ADHD symptoms, participants offered various responses. Some admitted addiction to prescription stimulants.
Others acknowledged recreational use in spite of their.

Im prescribed 360 Dexedrine a month. The past 2 months im lucky if they last a
week. So . . . you know youre an Adderall addict when you eat all of your own
prescription, as well as the scripts of about 7 other people.
> (RADioFluke, April 2003)

Recreational and necessary. Best of both worlds.
> (Got_me, December 2002)

Thus the diagnosis confused issues of abuse and recreational use for many. One participant reported that she no longer knew why she took Adderall. Another, who had reported purely recreational use in previous threads, became convinced that he suffered from ADHD.

> [Why do I take the Adderall] I just don't know anymore.
> (Corrupted, December 2002)

> [Now after taking the prescription] I really do think I have a mild form of ADHD.
> (AsianKing, January 2002)

Interestingly, the forum discussions let us see that not only are stimulant users concerned about getting drugs and managing their effects, but that they are working to define their own use. Ultimately the lines between recreational use, addiction, and medical need are blurry. Some participants were clearly and admittedly using medical need to legitimate recreational use and even addictive behavior. In this regard, obtaining a prescription stimulant as a result of a diagnosis may be the most dangerous route of access. Not only does it provide easy access to the drugs, but it may also encourage users to believe that their use is neither illicit nor harmful.

Conclusion

Compared to recent quantitative studies of prescription stimulant abuse, qualitative analysis identified many similar findings. Like that of Poulin (2001) the analysis indicates that drug diversion is common. The discussions also suggest that friendship networks, often developed at school, provide access to prescription stimulants for many users. Forum participants also cited reasons for use similar to those identified by Graff Low and Gendaszek (2002). They indicated that performance enhancement at school and work provided motivation for stimulant use. Stimulants were often cited as necessary for study and focus in class.

This study differs from previous work in that it examines the relationship between ADHD diagnoses and prescription stimulant abuse. The previous studies generally assume that ADHD diagnoses represent "real" and legitimate medical conditions suffered by respondents. They also assume that use by those with a prescription for the drug is not abusive. Forum discussion analyses offer evidence to the contrary. First, we see that prescriptions for pharmaceutical stimulants are often written as a result of drug-seeking behavior. Some participants cited detailed and specific methods of feigning ADHD in order to obtain drugs for recreational use.

Second, online discussions also reveal that those with potentially legitimate diagnoses still abuse the stimulants prescribed to them. Beyond providing channels for diversion, some abuse stimulants by taking a higher dosage than prescribed. The extent of nonmedical use among those diagnosed with ADHD cannot be determined through this study. However, it does highlight the necessity for future work that measures the degree to which those with ADHD abuse their drugs. Thus further quantitative work must separate the presence of a diagnosis from the nonmedical use of stimulants. It cannot assume that a prescription for a stimulant was legitimately obtained. Nor can it assume that those with ADHD do not abuse prescription stimulants.

Lastly, this study also suggests that abuse prevention efforts must address issues related to the inappropriate use of prescription medications. Several respondents with ADHD diagnoses questioned if they needed prescription stimulants. They also expressed some confusion regarding whether their use was recreational or medically necessary. These forum participants could benefit from efforts that reintroduce many to the symptoms of the disorder and directives regarding safe use. Similarly, others without ADHD began to think of their enjoyment of stimulants as evidence of the need for medical treatment. By obtaining a prescription, their justification for taking the drugs was confirmed by a doctor. This is perhaps the most critical finding of the paper—that doctor prescribed stimulants are not only fueling abuse, but that the process of diagnosis may confirm a felt "need" for stimulant abuse. Ultimately, drug-seekers that obtain a diagnosis may come to forget the dangers associated with stimulant abuse and may even come to define abuse as medically necessary.

References

Clure, C., Brady, K.T., Saladin, M.E., Johnson, D., Waid, R., and Rittenbury, M. "Attention-Deficit/Hyperactivity Disorder and Substance Use: Symptom Pattern and Drug Choice." *American Journal of Drug and Alcohol Abuse* 25(3), 1999: 441-48.

Coetzee, M., Kaminer, Y., and Morales, A. "Megadose Intranasal Methylphenidate (Ritalin) Abuse in Adult Attention Deficit Hyperactivity Disorder." *Substance Abuse* 23(3), 2002: 165-69.

Conrad, P. and Potter D. "From Hyperactive Children to ADHD Adults: Observations on the Expansion of Medical Categories." *Social Problems* 47(4), 2000: 559-82.

Drug Enforcement Administration. "Drug of Abuse." Arlington, VA: U.S. Department of Justice, 2003.

Finch, J. "Prescription Drug Abuse." *Substance Abuse* 20, 1993: 231-39.

Graff Low, K and A. Gendaszek. "Illicit use of psychostimulants among college students: a preliminary study." *Psychology, Health and Medicine*, 7(3), 2002: 002

Johnston, L. D., O'Malley, P. M., Bachman, J. G., and Schulenberg, J. E. "Monitoring the Future: national survey results on drug use, 1975-2004." Volume I, Secondary school students (NIH Publication No. 05-5727). Bethesda, MD: National Institute on Drug Abuse, 2005.

Klein-Schwartz, W. "Abuse and Toxicity of Methylpenidate." *Current Opinion in Pediatrics* 14, 2002: 219-23.

Klein-Schwartz, W. and H. McGrath. "Poison Centers' Experience with Methylphenidate Abuse in Pre-Teens and Adolescents." *Journal of the American Academy of Child and Adolescent Psychiatry* 42 (3), 2003: 288-94.

Leshner, A. "Understanding the Risks of Prescription Drug Abuse." *NIDA Notes* 16, 2001: 1-3. <http://www.nida.nih.gov/NIDA_Notes/NNVol16N3/DirRepVol16N3. html> (10 January, 2003).

Levin, F. and H. Kleber. "Attention-Deficit Hyperactivity Disorder and Substance Abuse: Relationships and Implications for Treatment." *Harvard Review of Psychiatry* 2, 1995: 246-58.

Musser, C. J., Ahmann, P.A., Theye, F. W., Mundt, P., Broste, S.K., Mueller-Rizner, N. "Stimulant Use and the Potential for Abuse in Wisconsin as Reported by School Administrators and Longitudinally Followed Children." *Journal of Developmental and Behavioral Pediatrics* 19(3), 1998: 187-92.

National Institute on Drug Abuse. "Prescription Drugs/Abuse and Addiction." *National Institute on Drug Abuse Research Report Series.* NIH Pub. No.01-4881. NIDA, NIH, DHHS, 2001.

Poulin, C. Medical and nonmedical stimulant use among adolescents: from sanctioned to unsanctioned use. *Canadian Medical Association Journal,* 165 (8), 2001: 1039-1044.

Robison, L. M., Sclar, D. A., Skaer, T. L., and Galin, R. S. "National Trends in the Prevalence of Attention-Deficit/Hyperactivity Disorder and the Prescribing of Methylphenidate Among School-Age Chidren: 1990-1995." *Clinical Pediatrics* 38, 1999: 209-17.

Safer, D. and Zito, J. "Adolescent stimulant use" [correspondence]. *Canadian Medical Association Journal,* 167(1), 2002: 15.

Stitzer M. and Walsh S. "Psychostimulant Abuse: The Case for Combined Behavioral and Pharmacological Treatments—The fourth generation of progress." *Pharmacology Biochemistry and Behavior,* 57(3), 1997: 457-70.

Substance Abuse and Mental Health Services Administration. "Nonmedical Use of Prescription-Type Drugs among Youths and Young Adults." *The NHSDA Report.* January 16, 2003.

———. "Overview of Findings from the 2004 National Survey on Drug Use and Health." Office of Applied Studies, NSDUH Series H-27, DHHS Publication No. SMA 05-4061. Rockville, MD. 2005a.

———. "Results from the 2004 National Survey on Drug Use and Health: National Findings." DHHS Publication No. SMA 05-4062, NSDUH Series H-28. Rockville, MD. 2005b.

———. "Stimulant Use, 2003." *The NHSDA Report.* February 4, 2005c.

———. "Nonmedical Use of Prescription-Type Drugs among Youths and Young Adults." *The NHSDA Report.* January 16, 2003.

Zielbauer, P. "New campus high: illicit prescription drugs." *New York Times,* March 24, 2000.

Chapter Nine

Illegal Behavior and Legal Speech: Internet Communities' Discourse about Drug Use

Sarah N. Gatson

The phrase "But, it's the *law*!" is a familiar expression. Among my students, this is so although I teach those who on the one hand are active in the movement to legalize same-sex marriage, and on the other, in the movement to re-criminalize abortion. Even as they are empirically engaged in explicit cultural/legal actions themselves, they often rapidly become uncomfortable at questioning the shape and content of their lives and their world. This is a situation present in the general population as well. There is often an assumption that the law is hard and fast. Although by looking at the statutory stage of law, we may see a sort of cumulative concretization, by only looking there, we miss the amorphous and ambiguous nature of the law, in both its central interpretations and its more marginal edges.

In the general public, legislative initiatives, and research agendas, the socio-political problem of drug use has been a hotly debated one. The uses of the Internet have been at least as ardently discussed. As I discussed in an earlier chapter, these two concerns have merged in the worry over the seeking and gaining of dangerous information, the issue of privacy, and both marginal and mainstream social movement politics. Of major concern to professionals in health care, policy-making, crime control, and education are spaces where youth in particular are exposed to communication about illegal and/or dangerous drug practices. These are spaces that encourage, teach, disseminate, and model such practices. Herein I seek to examine these everyday arenas of persons that are a nexus of drug use, discourse, community, and citizenship. I see these sites and the discourses and practices they model as having the potential to be equally about initiating youth into the responsible practice of community and citizenship. An organizing question for this analysis involves understanding both the Internet environment and the issue of drug use as situations of relative anomie. As Durkheim noted,

> for actions regarded as criminal to cease, the feelings that they offend would need to be found in each individual consciousness without exception and in the degree of strength requisite to counteract the opposing feelings. Even supposing that this condition could effectively be fulfilled, crime would not thereby disappear; it would merely change in form, for the very cause that made the well-springs of criminality to dry up would immediately open new ones (1982: 99).

Thus, I think we need to ask, are these folks Innovating or Rebelling? Are they possibly even Conforming to the democratic tradition (see Merton, 1938: 676)?

Any society has both some more-or-less stable structures, and some more-or-less boundary-pushing behavior aimed at (or affecting) those structures. How people go about living within those structures involves taking them for granted, as well as articulating explicit adherence and/or challenges to them. How explicit those understandings of the boundaries of one's social world are will vary depending on the particular state of flux of the society, as well as particular persons' positions within those structures. Durkheim noted, "duties are imposed upon us that we have not expressly wished. Yet it is through a voluntary act that they arose" (1984: 174). We may call this understanding of our social world "culturally oriented . . . response" (Merton, 1938: 672), "habitus" (Bourdieu, 1991: 50-52; 123; 176; see also Sumner, 1940: 3), "structuration" (Giddens, 1984), "rules of normatively appropriate . . . behavior" (Ellickson, 1991: 124), or indeed "legal consciousness" (Merry, 1990; 1995; Marshall and Barclay, 2000; Neilsen, 2000). All of these concepts together encompass a range of understandings about where in the continuum of making social control—and cultural life more generally—social actors may fall in any particular situation.

Sumner asserted that folkways, accreted and often unconscious habits of the past, were "converted into mores, and, by virtue of the philosophical and ethical element added to them, they win utility and importance and become the source of the science and the art of living" (1940: 3). However, Sumner admits that mores also "come down to us from the past" (1940: 76). Thus, folkways (norms) and mores (laws) must be seen to exist in more of a state of flux than in the analytically distinctive sense that understanding them as ideal types affords us. Ellickson wishes us to separate "ordinary human conduct . . . primary behavior" from "social control activity" which he posits may extend outward from primary behavior. He argues that we should distinguish from these levels of behavior in order to understand that "A guideline for human conduct is a rule only if the existence of the guideline actually influences the behavior either of those to whom it is addressed or of those who detect others breaching the guideline" (1991: 128; see also Becker, 1963: 147). This assertion clearly allows for a conflict vis-à-vis rules—not merely because some tend to be enforcers and others tend to be violators, but also because the substantive content of particular rules is differentially salient in a cultural sense, and at the individual level, as suggested by Durkheim.

In essence then, I want to call the bulk of the folks who make up the participants in the bulk of the websites I assessed in chapter 7, politically engaged deviants, or "counterpublics" (Travers, 2000: 147). They challenge the status quo in terms of both norms and laws, but they do it in common cultural discourse, emphasizing their central place in the arena of citizenship; as such, they are akin to Becker's moral entrepreneurs (1963: p.135; 147). Such politically engaged deviants are often at the forefront of social change, and they focus our attention upon the symbolic boundaries of our current consensus about what is right and wrong (see Morris, 1984: 50; 80). The field or scene of drug-using discourse thus shows us that society may be understood as an argument. Here we may find the intersection of the two rules in conflict—speak freely, and don't use drugs.

Bollinger asserts that the First Amendment generally, but particularly its free speech component "is . . . one of the nation's foremost normative and cultural symbols" (1992: 297; see also Demac and Downing, 1995). Free speech has several social functions/values in a democracy, among them wide-open debate, the search for truth, and individual freedom and autonomy (Ibid). In essence, it calls for responsible political engagement. Clearly, as it is this rule, rather than "don't use drugs," that has deeper cultural roots, as well as deeper roots in our legal and political sanctions, the rule (even the war) against drug use may be—and certainly has been—understood as open to debate.

Although so deeply embedded, we also know that free speech has a more recent formal interpretational history, beginning in 1919 with *Schenck v. United States* and *Abrams v. United States*. Both opinions were written by the same Justice—Oliver Wendall Holmes. Even coming from the mind of the same man, the idea of what the rule should be was different (Bollinger, 1992: 298). While generally Holmes's second major—and freer—interpretation has controlled, several restrictions play against the interpretation that content-neutral regulations of free speech should control our rules about it. In addition to usually being able to restrict the time, place, and manner of speech, other rules about speech are important—libel and slander, clear and present danger, bad tendency (is the speech likely to cause illegal action), distinctions between public and private arenas, fighting words, obscenity, and clear and present danger + imminence (Bollinger, 1992: 298-99; Rabban, 1992: 54). Clearly, my earlier approach to assessing the potential of informative/advocacy discourse regarding drugs online encompasses an approach to understanding how those websites may or may not be regulated. Here however, my point is not necessarily to emphasize how to stop their "bad tendencies," but rather to explore their discourse as marginalized—yet culturally core—politics.

Deviance, Politics, and the Law

As I noted above, the drug-using subculture—which in and of itself is a rather broad categorization—merges transgression and progression (see Gatson, chapter 7, this volume). That this transgression is ambiguous and contested is shown in recent polls of the U.S. population (Rohall, 2004: 60; Goode, 2004: 25). Discourse within the scene often acknowledges the risky nature of drug use, as well as argues for its potential to expand the mind, the self, and society; at the very least to liberate or lower a variety of socially-imposed (and thus ultimately arbitrary) constraints, enabling freedom (of thought, behavior, etc.) (see especially Gibson, chapter 5, this volume) there is an important thread of political discourse within both the rave and drug-taking scenes.[1] For example, among the many links on the website "The Halcyon Cosmopolitan Entertainment" (reciprocally linked at Erowid) is a link to The Dance Liberation Front (http://www.dlfnyc.com/). The DLF linked to an article about the FBI's labeling Reclaim the Streets.org a terrorist organization:

> The report reads in part: "Anarchists and extreme socialist groups—many of which, such as the Workers' World Party, Reclaim the Streets, and Carnival Against Capitalism—have an international presence and, at times, also represent a potential threat in the United States. For example, anarchists, operating individually and in groups, caused much of the damage during the 1999 World Trade Organization ministerial meeting in Seattle." The list also included "extreme fringes of animal rights, environmental, anti-nuclear, and other political and social movements" as well as the Animal Liberation Front (ALF) and the Earth Liberation Front (ELF). Getting included in such a list is always both a good and a bad sign: it means we're doing something right and are threatening the system, but it also vastly increases the likelihood of infiltration, frame-ups based on planted evidence, government-sponsored internal movement "splits," police use of deadly force, etc. Reclaim the Streets is actually more of a tactic than a movement or an organization. In 1996, activists in England decided to hold the first RTS "street party" by holding a day-time rave, complete with sound system, dancing, and party games, all with a political spin in a busy intersection. The party aimed to temporarily "reclaim" the street from cars and point out how capitalism and car culture deprive people of public space and opportunities for public festivals. The brilliant tactic rapidly caught on, and Reclaim the Streets street parties are now regularly carried out all over the globe. RTS goes beyond the limitations of the traditional "march and rally" protest by building coalition with the rave/dance/youth scene to create something that is disruptive and public like a protest, but that is also joyous, fun and beautiful like a party. Because it's fun and crosses over with the counter-culture, it's a lot easier for a street party to attract a large crowd. A street party can effectively shut down a business district, in a positive, militant yet non-threatening way. . . . it is particularly ironic and interesting that the FBI considers these dance-based parties as a "terrorist" threat. Where is the terror? Where is the violence? . . . we thought the FBI only got involved when there was something seriously illegal going on or people getting hurt. After wracking our brains, we figured it must be the video footage on Bay Area Reclaim the Streets's web page, which shows a car getting

flipped over during the first street party held in the USA on May 16, 1998. (http://xinet.com/ rts.) That street party was to protest the WTO, eighteen months before Seattle. Just for the FBI's benefit . . . that car was donated to us by a friend to help us block the street. We drove it into position in the middle of the streets, let the air out of the tires, and flipped it. It was just a prop. Kinda like art, ya know?[2]

As I discussed in chapter 7, about 17% of the Erowid linked pages contained some level of legal/political discourse, and this discourse was as often about the kind of political stance taken by the DLF and Reclaim the Streets as it was a straight-forward and unapologetic call to "legalize it."

These sorts of uses of the generally acceptable, positively sanctioned democratic process by such marginalized folks (or, at least, folks engaged in such marginalized behavior) present us with a particularly interesting instance of labeling deviants (see Becker, 1963; Kitsuse, 1964: 88; Davis, 1964). The DLF article's concern with inappropriate "branding" of legitimate protest behavior captures this interpretation well. When those engaged in deviant behavior resist the label of deviant with a response that is articulated as non-deviant, indeed crucial to the normal process of politics and making law, they highlight the complexity of labeling itself, as well as conflicts over situation, audience, and interest/reference group (Kitsuse, 1964: 99-100; Ellickson, 1991: 152-53). Davis reminds us that "it remains the more pervasive argument in a democracy to be able to claim that the social injury [making one a deviant] from which one suffers is in no way self-inflicted" (1964:119). However, "self-infliction," or individual choice and determination may be said to be one of the underlying tenets of the hegemonic political discourse of these online deviants. Thus, underlying much of the discourse under scrutiny herein is a notion of the right to be deviant, or rather to exist in a social environment where individual responsibility is presumed.

The Cultural Vibe of Erowid and its Audience

Erowid is an archive of a culture (ethno), specifically in the sense Zora Neale Hurston noted—the posts, the reciprocal links, are always available to others for further dissemination (Lionnet, 1991: 192; see also Gatson, 2003: 42; Gatson and Zweerink, 2004a: 194). An Internet site like Erowid is itself an ethnography in that it is representative of a dispersed social group with diffuse boundaries in which its members write their own narrative of their society, as it is, and how they wish it to be. In the site's privacy statement, we find an underlying orientation of individual freedom:

Many people ask whether simply visiting Erowid can be cause for concerns about privacy and security. One of the primary things to know is that we get nearly twenty thousand visitors a day (about five hundred thousand a month).

There is safety in numbers. Our visitors span the spectrum from users of psychoactives to law enforcement, educators, chemists, physicians, and parents. Visiting Erowid does not imply anything in particular about the visitor. . . . Because standard web logs track which pages are visited by each I.P. address, we have created several systems by which submissions (experience reports, images, etc) are considerably more difficult to track to a specific individual. For instance, when an image is submitted, the file write time is automatically changed so it will not be possible to determine who the submittor is by comparing the submission time to the web logs. All submissions can be made pseudononymously and contact information is not publicly displayed unless specifically requested by the submittor. . . . If you have previously submitted information to Erowid and would like to have your name or contact information removed from the submission, please let us know. Over the past six years we have incorporated several independent collections of information into our archives. Some of these archives may display names or contact information without the permission of the author. We are happy to anonymize submissions, though we will need to be able to verify that you are the original submittor.

At the bottom of this page are links to the Electronic Frontier Foundation, the Online Privacy Alliance, and Call for Action, all organizations that advocate the cultural notion of individual privacy and responsibility, and the free speech principle of a free search for information. As Fire and Earth, owners, webmasters, and founders of Erowid note, "Part of our work is to encourage those with hard-won bits and pieces of knowledge to record them and make them publicly available" (*Erowid Extracts*, 2002: 1).

Erowid as an organization, however, also acknowledges a communal responsibility; they worry about their community's reliance upon the information they have made so accessible, "It is somewhat disconcerting to hear—in email and in person—how much people rely on the site and how much trust is given to information we publish" (Ibid). That trust is there, as our survey bore out:

Erowid is key, the number one site, otherwise I see how consistently a particular statement comes up or do investigation to determine it myself."
White male, nineteen or younger

Erowid is the best example I've found. Not only do they have practical information on how to use drugs safely, but they have info on legalization/legal status, community, arts, and spirituality. . . . Erowid is also cool enough to have the nerve to talk freely about anything and everything. And reading other people's experiences before taking a drug helps you know what to expect.
White female, twenty to twenty-four

Sites that are not sponsored by the government. That have no hidden agenda about drugs being bad and that give intelligent information → Erowid.
White male, forty to forty-four

Sites that are reliable have gathered together information from multiple sources: texts, journals, users, etc. They also have a good reputation, especially

amongst those who use drugs responsibly. Erowid.org and its affiliates, are, in my opinion, the most reliable and accurate source of information for drugs that exists on the 'net. Their combination of information culled from reliable sources—even it contradicts other sources—and trip reports make for an excellent site.
White male, twenty to twenty-four

Well I find the most reliable sites to be non-profit organizations with no government affiliation. Government sites are going to be one-sided obviously and not tell you the WHOLE story. Sites such as bluelight.com and erowid.org are very informative because they collect data from MANY sources and compile it together.
American Indian or Alaskan, male, twenty to twenty-four

Whether the site looks professional or not—a site like "Yahooka.com" obviously isn't going to give a well round[ed] perspective of marijuana use. However through exploring erowid it's obvious that's a website dedicated to giving information about drugs, and not so much on "getting fucked up."
White male, ninteen or younger[3]

Fire and Earth Erowid themselves have demonstrated their commitment to bringing their marginalized cultural approach to drugs, community, religion, etc. straight to at least some of the powers that be. In June 2002, they were invited to participate in a small working conference at NIH/NIDA, a meeting our project participated in as well (Fire Erowid, 2002: 2).

Our survey also asked respondents, "Why do you participate in discussion boards or chat rooms? What do you enjoy talking about the most?" There were 249/831 (30%) instances where respondents used "info/information" and "knowledge" as keywords to discuss their reasons for participating in CMC, as well as 299/831 (36%) instances of using "connect" "similar" and/or "talk." These comments often had the flavor of the underlying Erowid cultural values I discussed above,

White male, twenty-five to twenty-nine: The amount of people that are interested in these subjects at the level that we are is low. We are spread out geographically. I am able to connect and discuss a topic with others that is taboo. There is no place that educated middle of the road working class people hang out at where they freely discuss these topics.

White male, ninteen or younger: You can't talk about drugs freely in day to day life without getting weird looks from people and the police at your door. Discussion boards give me a chance to ask questions I need answered, or just to share a good time I had without having to worry about who's looking at it.

White male, twenty to twenty-four: I enjoy the candid discussion and the fact that people can talk about these things without being locked up or persecuted. I enjoy talking about experiences and learning from others experiences to make my own more safe/enjoyable.

White male, thirty to thirty-four: Initially, it was to find information on drugs, then to share the knowledge I had gained. I'm pretty much bored with talking about drugs now, though I do occasionally. I'm basically there for the community (Bluelight). People know me, we can have a serious chat about anything, or a joke. Because of the board I've been able to meet people all over the world—I could go almost anywhere and someone would want to meet me—it's a community I'm proud to be a part of.

White female, twenty-five to twenty-nine: The reason I started participating in discussion boards was to find out more about ecstasy. I've since become less interested in talking about drugs and more interested in talking about music. I participate because I've met a lot of people online and it's now become a way for me to communicate and socialise with friends while I'm at work and at home.

White male, twenty to twenty-four: Nothing really, I just lurk. Its a form of entertainment and it sure beats watching commercial TV. I can see if other people out there have similar drug and life views and experiences.

Finally, we asked the question, "If you could sit down with us and tell us anything that you think we should know about recreational drug use, music and raves, online communities, or the government's policies toward drugs, what would you say?" These answers were more explicitly on the libertarian end of the free speech realm—arguing for free thought, free speech, free information, and the free action of the harm reduction discourse,

White male, twenty-five to twenty-nine: Any consenting adult should legally able to ingest whatever they want. Raves can be, and often are, positive, life-affirming experiences, with or without drugs. Raves are not for the purposes of drug use, rather it's usually the other way around. People take drugs to enhance their music and dancing experiences. The Internet has been a dramatic force for drug-policy reform, allowing the public information on drugs they wouldn't normally have. Recreational drug use, as opposed to addictive drug behavior, is rarely harmful in the short term and may not be very harmful in the longterm, it really depends on each individual user. Given honest information, users tend to make the best decisions for themselves.

White male, twenty-five to twenty-nine: As for the internet, people were taking drugs long before there were websites about it, and I think though some will take more/different drugs as a result of drug websites, I think they swing the balance in favor of use over abuse.

White male, twenty to twenty-four: The Center for Cognitive Liberty and Ethics (www.http://www.cognitiveliberty.org) has the right idea, that liberty, privacy, and self-determination over one's own intellect is essential to our most cherished freedoms. i would think surely in such a "civilized" world, we would be allowed to do our own thinking, without somebody telling us how we should think and live and run our lives. as for music and raves, since when is art and public gathering for appreciating it a crime? music is art, and people do gather to enjoy it. making such things illegal does nothing but create crime.

Thus, Erowid clearly articulates an ethic in this cultural field, and explicitly uses a new(ish) technology in order to facilitate that ethic.

Methods and Data

In doing online research, there are a variety of behaviors to assess, and a variety of ways to describe what one must do to make those assessments. Online-based research may be the participant observation of live CMC, it may be participant observation in, or the reading of, synchronic and diachronic CMC, and it may be reading of existing text/graphics on a variety of websites. This depends on the kind of website, whether a community is involved, and whether the communication patterns of the producers of the text(s) involve talking back to their audience, or allowing for the audience to talk to one another. Here all of the data is text and/or graphic images (still or moving) produced by more or less distant persons, and it is meant to be responded to in some way. It is conversational text, available for public consumption that reflects network development if not always community development (see Gatson, chapter 7, this volume). It often rises to the level of general cultural discourse.

I have long focused my methodology upon documents, official and unofficial, public and private, and junctures between. My own first ethnographic sites involved communities originating in computer-mediated communication areas of the Internet, thus I have moved into an area where ethnography is largely text-based (Gatson and Zweerink, 2004b: 1-28). Dictionary definitions of ethnography explicitly refer to "the descriptive anthropology of technologically primitive societies," although once one moves down to ethnomusicology, it is merely "the study . . . of different cultures," not necessarily "primitive" ones (*American Heritage Dictionary*, 1991: 467). The literal meaning of ethnography is merely that of writing about some group—although "ethno" means race or people, "race" or "people" can mean simply some group of persons with more or less permeable and explicit boundaries. Lionnet uses it to mean culture (1991: 188), and in Dorothy Smith's notion of "writing the social," she suggests that the "multi-voiced society" is always converted into texts by sociologists (1990: 137-138). In looking at the central node that is Erowid, we are—at least in part—seeing a dialogic "interchange between the state and its people," and are thus engaged in *structurated ethnography*—one that combines the extended case and multi-sited methods (see Gatson, chapter 7, this volume; D. Smith, 1990: 138; Burawoy, 1991; 2000; Marcus, 2000).[4] For this chapter, I took the fifty-two sites assessed in chapter 7 as a subset of the Erowid audience, and read them similarly to Erowid itself—nodes more or less networked into a scene in which drug use is one component of a deep political discussion.

Discourse—Text Versus Behavior

In moving from garnering a general sense of Erowid sites, and assessing the "bad tendencies" of the 52 CMC sites I investigated, between April 23, 2004 and January 15, 2005, I re-read the fifty-two sites (live, when accessible, and my saved files of older perusals) for political/legal discourse specifically, often finding it in conjunction with discourse about drugs, but not necessarily. I coded for discussions of 1) discourse on legal (or over-the-counter) drugs, including alcohol, tobacco, and caffeine (17%), 2) discourse on illegal drugs (62%), 3) political-legal discourse, including definitions, positions, philosophy and partisan (58%), 4) calls for particular political actions (such as the DLF article discussed above) (23%), 5) positive information about drug use, such as trip reports and advocacy in either a recreational, spiritual, or medical sense (54%), and 6) negative information about drug use, including cautionary tales (25%). These percentages do not total 100 because as I read the sites as wholes, as one might peruse a book or stack of documents and note the variety of themes, voices, and subject matter, each site usually contained more than one kind of discourse. That the majority of these sites had political discourse should not be surprising, given that they are self-selected reciprocal links to an online venue that casts drug use as part and parcel of our political—as well as medical, religious, leisure, and legal—decision-making landscape.

Taking the Big Fun website that I discussed in chapter 7 as our first example, there is a plethora of discourse on legal drugs, positive information, calls to try various drugs, with particular detail on Tussin, and some marginal or illegal (absinthe; LSD; ketamine; Ecstasy) drugs. Additionally, there is a considerable amount of reference to political positions and activities involving the Big Fun folks, ranging from fascism, anarchy, racism (discussed as both "redneck" and neo-Nazi skinhead). However, as the Glossary was written by The Gus, his underlying theme of individualism is the most predominant,

> To clear up any confusion on this point, I'd like to state that I do not identify with any one particular movement or subculture. I have drifted among several (everything between computer nerd, redneck hesher, hippie, environmental radical, punk rocker, art fag, and even goth to an extent) without ever being completely absorbed or finding complete satisfaction. The defects in most subcultures are usually immediately apparent to me and thus easily avoided. The most visible cultural influence at Big Fun was obviously that of punk rock, the culture with which I have been most consistently associated (at least since 1987). Be that as it may, however, I do not listen to very much punk rock music and am unlikely to start wearing spikes even though many of my friends do both. I do pretty much what I want to do and get from punk rock what I need from punk rock.

From Big Fun, one can go to The Gus's current and up-to-date weblog, where his political commentary is even more extensive and explicit: "You have to love the corporation-crushing democratization possible with the World Wide Web. . .

. At this time in American history, for me to post that page, satire or not, is to break a taboo even more sacred and arbitrary than the one attached to childhood sexuality."[5] Although not predominant, there were some examples like The Gus—one individual's take on politics, demonstrating an individualist, yet communitarian theme. One of these is Jessamyn (Case #33) another person with an explicitly public and on/offline linked Internet identity,

> Q. Why do you call yourself an anarchist?
> A. I do it less and less lately mainly because I am tired of associations with people who are really violent or really knee-jerk reactionary. I think of myself more as anti-capitalist, though I agree with many of the anarchist tenets. Loosely, I believe people would be better off with less governmental interference, I'm against hierarchies, and I believe that humans have a responsibility to look after each other and take care of one another so that everyone's strengths and creativity are utilized and maximized. I also think everyone should be actively working towards this, not just paying lip service towards the idea of the good life. To this end, I think we should shop less and create more, complain less and do more, and argue less and talk more. I'm not good with slogans, feel free to drop me an email if you'd like a more in-depth explanation of all this. . . . I try to tell the truth, to lead an ethical life and live low on the food chain, buying very little and trying to shop responsibly when I have to shop. I try to entertain myself easily, visit as many of my friends as possible, travel frequently and never ever say I'm bored. Talking about politics, particularly interpersonal politics and dynamics is very interesting to me. I think the more we are prepared to back up our beliefs, the more we reflect on them and inspect them for flaws. (From Jessamyn's FAQ page)[6]

Far more common than these folks though were the sites like Cannibis.com (Case #2), where explicit group political action, often vis-à-vis drugs, was the norm:

> The first [lesson in the history of drug criminalization] is that hate does not pay. It is ironic that the racism of the American people would end up hurting them this way—a sort of divine justice if you will. Because Americans were blinded by fear, hatred, and intolerance of other races, they allowed a prosperous future to slip between their fingers. Another thing this whole history tells us is that Americans need to take Democracy more seriously. If they had devoted more of their time to informing themselves about the world around them, they would have known what the real issues were. Instead they read the tabloids—look where that has gotten us. Finally, now that we have put marijuana prohibition into historical context, we can see clearly that it had nothing to do with public safety, or national security, or what have you. By all rights, marijuana should not have been made illegal in the first place. If today prohibition still has no rational basis to stand on, then let us repeal it. ("Well, why aren't we using hemp then?" Cannabis FAQ part II)

The creators—and often the users—of these sites have sophisticated enough legal consciousness to track the history of particular laws, and to put their own spin on what that history means for the legal present.

The participant audience of Cannabis.com often posts their further interpretation of this organization/website's political stance:

> **Message #268** posted by corruptedyouth. (Info) February 08, 2003 03:42:21 ET
> I was talking to my friend about this idea last night, before I read the post about the organized smoking protest. We both thought this would be an EXCELLENT idea. I have some improvements to make . . . Most of the rallies should take place in Washington, D.C., for obvious reasons, for example on the steps of Congress. I understand that not everyone would be able to miss work (work?) and go to D.C. I like the idea about the parks, but if we were to get some SERIOUS publicity, and that is what we really need, we are going to need a MASSIVE gathering. . . . I was also thinking that some insane cops could go as far as trying to incarcerate us and prosecute us as (the dreaded T-word) TERRORISTS! That would get a lot of publicity. You have to understand that the public mostly supports the government from what they see through the boobtube. Whatever the government tells them, is what is important. The government (minus the FCC) can not control the media under the right of Freedom of the Press. . . . BTW. To prevent ultra-prosecution and basically personal money loss, only bring 2-3 joints to the organized smokeouts. The cops are going to take them. Why waste your money on a half-o when the cops are going to take it anyway (and smoke it!)?
> Stay motivated,
> Fight for MMJ and legality,
>
> NITSUJ
> 2/8/03
> (one more thing. i heard somebody talking about erecting a memorial to commemorate the people who have died and those who have died in prison for drug charges once it is legalized. AND IT WILL BE!. This would give us a good image as real-Americans rather than skeevy stoners)(BTW stop the stupid NPDFA ads on tv. i have a bump the size of texas from when i found out that marijuana causes teen pregnancies. it wasnt the ad that gave me the bump, it was from when i slammed my head into my school desk 3 hard times in anger.) I dont smoke anymore. I am for the cause. And i am in high school. **I will be 18 next presidential election so Bush better choose his words wisely. support the green party.**
> thanks for reading all that

> **Message #234** posted by Holden99 (Info) September 25, 2002 22:55:33 ET
> I'd love to organzie a get together of supporters in Iowa or surrounding states to voice and live what we believe—My vision is something very small at first, just a gathering of a few trusted, known and politically active people to spend some time to brainstorm, empower and maybe get a little high if that's your thing. Eventually, it it succeeds we can slowly make it bigger and bigger and slowly adapt and influence the mainstream. America isn't ready yet but grassroots with a vision can move mountains. Email if you're interested or have ideas.

Message #115 posted by SweetLiberty2 (Info) January 15, 2002 23:56:32 ET
As part of a Nationwide Bill of Rights Campaign, a rally focusing on how our freedoms have been eroded over the past 2 centuries by politicians and agencies acting outside the law, will be held this Saturday (Jan. 19) in Denver. Starting at High Noon we will be gathering in Lincoln Park which is across the street from the West steps of the state Capitol.

Topics will include the new "Patriot Act," the illegal War on Drugs, unlawful searches and arrests by Law Enforcement Officials, and more . . . Speakers will include:
Mr. Rick Stanley, Libertarian candidate for U.S. Senate 2002—Colorado *** Rick got arrested at last month's rally for an act of civil disobedience . . . will he manage to go home after the rally Saturday night, or will he spend another 27 hours in jail for another act of civil disobedience? Come and find out!***
For more on Rick's arrest please see:
 http://www.stanley2002.org/denvervsconstitution.htm
Mr. James Vance, Libertarian candidate for Colorado Governor
http://www.jamesvance.com
Mr. Ralph Shnelvar, Libertarian candidate for Colorado Governor
http://www.lpcolorado.org
Mr. David Segal, who Denverites will remember for organizing some very successful Bill of Rights rallies in years past.
Michelle Konieczny who will speak about Abortion and the Ninth Amendment as Saturday is the anniversary of Roe vs. Wade.
and others . . .
Come stand up for your rights. A large crowd will generate media attention which could finally get the truth out to the public about the state of our nation. Come hear what your government is REALLY up to, what the media won't tell you, and what you can do about it.
See you there!
Denver Bill of Rights Rally
Saturday, January 19, 2002—High Noon
in Lincoln Park across the street from the West Steps of the state Capitol.
****
****
For Freedom,
SweetLiberty2
In Colorado?
Why vote for one of the Drug Warriors when you can vote for a man who
 knows the War on Drugs is the real crime?
Vote for Rick Stanley, Libertarian for U.S. Senate 2002!
http://www.stanley2002.org
"This time make your vote count"
Nationwide Bill of Rights Campaign—More Information.

With posts of this nature going back through at least 2001 and up until the present—such as Libertarian Toker's post about Libertarian Party presidential candidate Michael Badnarik's pro-hemp legalization stance and a link to the Bad-

narik/Campagna campaign website (6/14/04)—Cannabis.com, like Ya-
hooka.com, 420.com (the presence of NORML online), Chicago Ravers (Case
#3) and The Deoxyribonucleic Hyperdimension (Case #24), are evidence of an
articulate and politically engaged scene.[7]

The sites also often demonstrate community standards on speech, as does
Chicago Ravers' statement on the ground rules in their Substances forum:

> no drama, flaming, or "cyber-grab ass." In other words if you come to this fo-
> rum be prepared to debate like an intelligent adult. Childish behavour will not
> be tolerated. Come here to make a point and/or share your expereinces. Don't
> just post "drugs are bad" "drug users are losers" etc. No spreading misinforma-
> tion or dangerous information. . . . Come equiped with the facts or don't come
> at all. Stop spewing the rumors your baby's momma's mother's ex-husband's
> cousin from down the street saw on the Ecstasy episode of Oprah. Also, even if
> you'r just trying to be funny, don't go posting tips for huffing gasoline or how
> to smoke rat poison. It hard enough to tell when someone's telling a joke in cy-
> berspace. Beleive me, some one might just be stupid enough to try it. No solic-
> iting of illicit substances. Don't use this forum to try to hook up. Do that on
> your own time. THis forum isn't here to be your supplier. Any failure to obey
> the above rules will result in your posts being edited, closed and or deleted.
> You may also be suspended or banned from the forum or board. I reserve the
> right to change the rules as I see fit and may make additions and/or subtrac-
> tions. Check back every so often to make sure you comply.[8]

Evidence of a more general legal consciousness are the disclaimer notices that
often serve as textual gates to site content. Psiconautas (Case #13) provides a
typical example, in both Spanish and English:

> No somos responsables del uso que hagas con la información publicada en este
> portal, así como tampoco del contenido de otras páginas web a las cuales es po-
> sible acceder através de nosotros. . . . ESTOS ARCHIVOS SE
> PROPORCIONAN SÓLO COMO MATERIAL EDUCATIVO E
> INFORMATIVO Y ESTÁN DESTINADOS A PERSONAS ADULTAS QUE
> HAN ALCANZADO LA MAYORÍA DE EDAD LEGAL.

as does Drugs-plaza (Case #25), "All information on this website is for educational purposes only and is not intended to condone or promote or incite the use
of illegal or controlled substances," and Herb and Psychedelic QandA (Case #5):

> (2) THIS FORUM IS FOR THE LEGAL DISCUSSION OF THE HEALTH,
> SPIRITUAL, CULTURAL, RELIGIOUS, AND LEGAL ISSUES
> ASSOCIATED WITH THESE SUBSTANCES, AND THEIR USE.
> I AGREE THAT I AM OVER 18 AND I AM MATURE ENOUGH TO
> DISCUSS MARIJUANA AND PSYCHEDELICS IN AN OPEN FORUM
> SETTING. I AM NOT OFFENDED BY MARIJUANA USE OR THE USE
> OF PSYCHEDELICS. I ALSO AGREE THAT I CAN FOLLOW THE
> LISTED RULES. FURTHER, I UNDERSTAND THAT EVERYTHING I
> READ ON THIS FORUM MUST BE REGARDED AS FICTION, AND

THAT THE ONLY PLACE TO RECEIVE A VALID ANSWER TO A MEDICAL QUESTION IS FROM A MEDICAL DOCTOR. CLICKING ON THIS LINK CONSTITUTES AN AGREEMENT TO ALL THE LISTED TERMS.

In this subset of fifty-two sites (from the original 282/360 reciprocal links) legal disclaimers are not that prevalent (about 10% of sites contain a disclaimer of some kind), demonstrating some understanding of the formal rules of disclaiming responsibility within free speech doctrines.[9] We may understand such disclaimers as attempts to get around bad tendency. In their advocacy of individual responsibility, their claims to providing information, education, or entertainment only, and their giving notice to minors to stay away, these site owners attempt to place their information and CMC-hosting as places for considered and deliberative information-gathering and legitimate political discourse. At once, they claim no responsibility for the uses others may make of the information they provide, or for which they provide a venue, and also make the claim that theirs is a place for responsibility—the responsibility of democratic individuals who can and must engage in speech. When not explicitly using a disclaimer to make this last point, a site's owner or manager may instead make a legitimacy claim in another way, "Our goal at Ya-Hooka is to unite the marijuana Internet community, and to become an online demonstration that we are in every neighborhood, state and country, of every vocation, age, social group and color" (Case #21).

Discussion and conclusion: Bounded behavior, unbounded discourse?

As I noted in chapter 7, it is extremely difficult to know whether, in a bad tendency sense, the websites I have explored are the gateways to "gateway drugs." Vis-à-vis a host of online and offline behavior, bad tendency and free speech/privacy debates more generally are part of our ongoing cultural contest (Reed, 2000; Stembridge, 2000; Finkelman, 1999; Lee, 2001; Krotoszynski, 2003; Komasura, 2002; Delgado, 1995; Cenite, 2004; Ross, 2002; Volokh, 2000; and Brenner, 2002). It is less difficult to know whether these websites are reflective of an ongoing political/legal subculture that seeks to be taken seriously as political actors. They are engaged, ethnomethodologically, in helping to make up our current dialogue about these issues, and they are doing it with savvy, with politically powerful discourse, even if they themselves are not central political actors. As Lee argues,

> The initial situation in which individuals find themselves is the force field of language (representations, words, names); individuals are always already subject to the effects of this field (symbolic power). Individuals and groups also have a potential capacity to resignify themselves via the creation of new words and representations or through the revaluation of existing ones. The crucial

questions that emerge from this initial situation are whether and how the effects of this force field and the unequal capacity for resignification and/or re-
valuation can be intermediated by democratic procedures. (Lee, 2001: 868-69)

Thus, there is an ongoing "force field" of a rich discursive tradition over the
legal boundaries of questionable, deviant, and criminal behaviors (see also Gatson, 2004a; 2004b).

The information and interactive website network in which the
rave/dance/drug-using scene is at least in part embedded is also nested within a
larger and ongoing debate about use and abuse, and the legalization of particular
substances (See Conrad, Pinchbeck, Redden, Zimmer and Morgan, generally).
As well, this discourse is linked to the ongoing debates about other legal issues—such as free speech, privacy, and democracy—on the Internet and in society generally (See Demac and Downing, 1995; Demac and Sung, 1995; Etzioni,
1992, 1998a, 1998b; Gandy, 1995; Godwin, 1998; Goldring, 1997; Gurak, 1997;
Hickok, 1991; Lessig, 1999; Lipshulz, 2000; Loader and Thomas, 2000;
Mackinnon, 1998; Mayer-Schonberger and Foster, 1997; Reidenberg; 1997;
Rheingold, 1993; Riley, 1996; Schiller, 1995; Sclove, 1995; Smith, McLaughlin,
and Osbourne, 1998; Sobel, 1999; Sutton, 1996; Walzer, 1983, 1992, 1995;
Wallace and Mangan, 1997).

There are several policy implications, at several levels of government, inherent in recognizing that the nexus under investigation herein is embedded in
these ongoing discursive (legal/political) traditions. NIDA itself has recognized
some of them, extending serious invitations to organizations like Erowid to participate in policy discussions: "It was a pleasure to have a chance to share ideas
and debate viewpoints with thoughtful researchers who normally live on the
other side of the invisible fence of politics" (Fire Erowid, 2002: 2; see also Earth
and Fire Erowid, 2002: 6-7).

Health policy
With the Internet, there is an easy and relatively cheap way to set up discussion
and information websites like DanceSafe and Bluelight. NIDA/NIH could either
directly run and control these, or indeed there could be more official and open
collaboration between NIDA/NIH and various arms of the "harm reduction"
movement. There is a need to recognize that the promulgation of exaggerated
hysterical ad campaigns, and easily contestable (and in some cases, disprovable)
statistical data both do little to stop harmful drug use, and may do much to increase cynicism and distrust of health care authority figures. A realistic attitude
both towards current drug use, as well as toward the fact that the "youth" population is one that is ever emerging and aging-out into legal decision-making enti-
ties—adults, in other words—connects directly to more explicitly legal and political policies discussed immediately below (see Steinberg and Carnell, 1999;
Steinberg, 2000).

Local policy
Steinberg has argued that the goal of pinning down the exact age where adolescents enter adulthood, in the sense of emotional, cognitive, and moral responsibility, is nearly unattainable (2000). In contrast to the dichotomy of adult versus youth, he proposes a three-tiered approach based on age and development to determine legal responsibility (Ibid). Steinberg's focus is explicitly upon the juvenile justice system and the problem of trying children/youth as adults, his arguments regarding the amorphous and somewhat arbitrary nature of determining the threshold(s) of adulthood are useful here. Local approaches to dealing with the (perceived) problem of extant and/or increased childhood and adolescent drug use and abuse could be developed that take advantage of the extant models of Internet discussion groups as well as the extant models of after school programs in which youth engage in both peer-directed and adult-supervised discussions. These after school/extracurricular discussion forums could be designed after the model of DanceSafe and Students for a Sensible Drug Policy, with locally-decided criteria and moderators chosen from local meetings among parents, teachers, and youth themselves. Funding for such projects could use local Community Development Corporation networks, and Community Development Block Grant funds, among other local-national resources.

Rather than focusing on an "abstinence-only" policy, these local initiatives, supported by state and federal resources, could focus on giving accurate information to youth, creating a safe arena (in terms of information accuracy, confidence in the legitimate authority of parents and teachers, and using the faceless aspect of the Internet to encourage youth to give voice to their real concerns), and developing the emotional, cognitive, and moral responsibility aspects of attaining adulthood. Children are members of communities and have varying levels of access to citizenship. They are involuntary members of communities who are growing into decision-making and purposive citizens. They will take their places in society as fully legal members. It is the job of the current generations of adults to ensure that they have a voice as youth, and that their future responsibilities as both informed community members and leaders have a robust basis in conscious, engaged, and free community and citizenship participation (Gatson, 2004b).

Erowid has already begun such a local policy approach, using a grant to start its Families and Psychoactives Vault, where they assert that "it is rare for parents or other adults to actively engage their younger relatives and children on this topic" (Sophie, Earth, and Fire Erowid, 2002: 10). This concern is reflected in online venues. At the Hippyland site (Case #8), there was a thread called, "'its more harmful than we all thought' ****what do you think>>>" regarding the latest round of Public Service Announcements regarding the harmfulness of marijuana,

> **djordje says:** oh, i think you're talking about the commercial with the kid cutting open the cigarettes and rolling them in a new paper. yeah, those comercials are hillarious. well, yeah, ounce for ounce, weed has more tar(resin) and junk in

it than cigs, but a lot of smokers smoke a pack a day or more. there are what, 20 cigs in a pack, so thats like smoking 5 joints a day, every day. i don't know many kids that even smoke that much herb. the commercial is kinda right, but it doesn't quite tell a full truth, more like a half truth.

GanjaFairy says: ok so four cigs = 1 j how many cigs are in a pack? 20? i bet more people smoke a pack of cigarettes a day than five joints. thats about a quarter oz a day. (correct me if im wrong) even i, a daily smoker, do not burn through a quarter a day by myself. also, pot is nontoxic, unlike all the additives in most cigarettes(only additive free ones i know of are american spirits). pot doesnt cause cancer. pot doesnt cause emphysema. pot can be used to treat a variety of different illnesses. according to research reports ive read online, regular marijuana smokers are pretty healthy. i dont know if links are allowed in this forum, but this page has a lot of good information http://www.erowid. org/plants/cannabis/cannabis.shtml erowid has a lot of good info actually

KieF says: quote:
and as for that other ad in the campaign, the one about how such and such per-centage of kids tested for drugs at the scene of accidents tested positive for marijuana, that ad's just too fucking broad. for example, were those kids the ones who CAUSED the accidents?? what OTHER drugs were present in their systems?? i bet most of them were caused by drunk jackasses with no right to even be near a car. but no, they want you to think its the stoner's fault cuz pot's illegal and alcohol isn't.

yea . . . I heard that statistic in Health class in 12th grade and I thought it was BS. . . I talked to my teacher after and he said even he thought all those ads are overdone and don't make good points, and that he just had to follow what the ciriculum gives him, and facts out of the text book. (4/21/03; 18 replies, 336 views).

State policy

There is already a conflict between states within the United States over drug policies. Nevada has essentially decriminalized possession of less than one ounce of marijuana, thus making personal use—recreational or medical— without punishment feasible. California has an ongoing conflict with the federal government over their liberal medical marijuana practices. Other states may conceivably move in these directions, as Alaska has attempted to recently. In light of the recognition of the amorphous and arbitrary standards towards both youth and drug use historically and contemporaneously, perhaps lawmakers and citizens generally should be supported in forming formal Internet discussion groups like those outlined in policy sections A and B above. Let us take the po-tential for interaction through CMC seriously, and position health experts, policy makers, and the citizenry in Internet NGOs.

Federal policy

Federal policy operates both nationally, arbitrating between and among states, and internationally, emphasizing borders between the U.S. and the rest of the world. See policy section immediately above.

International policy
As with the conflict among and between the states, there is a conflict between Canada and the U.S. regarding the concepts of drug use and drug abuse. While national borders are important, the reality of an international debate about these issues, as well as the interconnected nature of nations and their economies (legitimate and underground) is just as real. The recent Canadian Senate study recognizes this situation explicitly (Cannabis: Our Position for a Canadian Public Policy; see especially chapter 4 of the report; see also Berman, 2003). We live in a world that houses the extremely liberal policies of The Netherlands, as well as the extreme violence of the illegal drug empires of Columbia. Both of these political-economic situations are intertwined with our own internal economies and politics, from the local to the federal. In sum, for policy implications in these final three sections, I advocate recognizing the existence of fairly well organized pockets of resistance to the "abstinence-only"/"just say no" approach to dealing with illegal drug use.

Our most realistic policy suggestion for these levels of government involve taking real advantage of both the multi-user communication and community functions of the Internet, and its natural surveillance capabilities. We mean to take advantage of the latter not in a "Big Brother" sense, but in contrast, in a participant observation sense, listening to the citizenry at large. Can (Should?) we really use the Internet to promote liberal democratic, or libertarian, values in this way? Can (Should?) we make "democratic participation [if not] participatory democracy" rather than totalitarian understandings of political participation for youth (and the rest of us) more likely? (see Rodman, 2003: 39; Froomkin, 1997: 155-56). Are the risks inherent in drug use worth tolerating for the realization of the values inherent in the also risky use of free speech? As Rodman asserts, these are "messy and complicated" questions that are going to have different answers across "particular contexts" (Rodman, 2003: 39).

Notes

1. I consider this to be serious politics, not the "sideshow" that Goode asserts it to be (2004: 25). Arguments advanced online by political actors are also made by those more likely to be understood as seriously engaged political actors, such as Executive Vice Presisdent of the Cato Institute David Boaz, as evidenced by his testimony before the House Subcommittee on Criminal Justice, Drug Policy, and Human Resources (http://www.cato.org/dailys/06-16-99.html).

2. The website hosting this post was not part of the database of assessed websites in Gatson, 2004, as neither Halcyon nor DLF had significant CMC, and as of April 28, 2003, the DLF website got a "page cannot be found" error message when attempts were made to link back to it. However, "The goal of the Dance Liberation Front is to eradicate New York City cabaret laws, which prevent dancing in unlicensed venues. Norman Siegal, president of the NYCLU [New York City chapter of the ACLU], is the legal repre-

sentative for the DFL." Thus, this particular submovement of the scene likely continues to have an offline presence. The Northwest Late Night Coalition with a similar political profile as the DLF, linked from Case #15, is in the database.

3. Fire and Earth constructed and published a survey on the perception of risk of Ecstasy, as well as asking respondents to rank the relative credibility of eleven different information sources. Erowid and DanceSafe were ranked the highest (2002: 9).

4. I mean structurated in Giddens's sense of structuration—the way in which the making up of the structure and the individual/group are simultaneous and mutual (Giddens, 1984). This form of ethnography may also be discussed as ethnomethodology, or be simply several levels of fieldwork—in that our project engaged in survey methods, participant observation, and the analysis of documents (of both a primary and secondary kind).

5. As I noted in chapter 7, I assessed the primary reciprocal links, and not generally the secondary links. The Gus is an exception, in that he maintained Big Fun's history online, and that segued into his own weblog; they were assessed as essentially one case. There were a few others that I assessed in this way, for example, Case #17, where the initial link passed me onto the principally-coded link, which site owner passed me on to the official website of the Om Festival of Canada, thus allowing for a coding of how linked up online and offline the primary site was, and how much potential it had for enabling on/offline linking.

6. FAQ is the acronym for "Frequently Asked Questions," and is usually a webpage or list on a webpage answering such questions about a site or topic. When I say public and on/offline connected in Jessamyn's case, it is in the sense that she gives extensive detail on her offline identity and location (her geographical address is freely accessible on her Visit Page, and she has an open-house policy on visiting her at her home—just drop her an email, begin a conversation, and she is willing to have you visit. She's been doing this for a few years, calling it "Bed and Breakfast style w/o the high prices!"

7. Because these sites and others like them typically support multiple topic areas (threads), I must also point out that such serious political posts and discussions occur alongside—sometimes within the same thread—discussions about meeting up for partying and/or romance, arguments, flame wars, and less deliberative and more insulting conversations.

8. For extensive discussions on online community standards and enforcement thereof, see Travers, 2000: 53-72, and Gatson and Zweerink, 2004b: 141-188.

9. In this sense of legal consciousness, more sites used standard copyright and fair use language—they were more concerned with intellectual property than with bad tendency. In the original set of reciprocal links legal disclaimers were even less prevalent (about 4% (242)/3%(360))

References

Cases Cited
Abrams v. United States, 250 U.S. 616 (1919).
Schenck v. United States, 249 U.S. 47 (1919).

Secondary Sources
American Heritage Dictionary. Boston: Houghton Mifflin Co., 1991.
Becker, Howard S. *The Outsiders*. New York: Free Press, 1963.

Berman, Paul Schiff. "The Internet, community definition, and the social meaning of legal jurisdiction." Pp. 19-82 in *Virtual Publics: Policy and community in an electronic age*, edited by Kolko, Beth E., (New York: Columbia University Press), 2003.

Bollinger, Lee. "First Amendment." In *The Oxford Companion to the Supreme Court of the United States*. Oxford: Oxford University Press, 1992, 297-99.

Bourdieu, Pierre. 1991. *Language and Symbolic Power*. Ed. John B. Thompson. Trans. Gino Raymond and Matthew Adamson. Cambridge, MA: Harvard University Press, 1991 [1982].

Burawoy, Michael. *Ethnography Unbound: Power and Resistance in the Modern Metropolis*. Berkeley: University of California Press, 1991.

Burawoy, Michael. *Global Ethnography: Forces, Connections, and Imaginations in a Postmodern World*. Berkeley: University of California Press, 2000.

Cenite, Mark. "Federalizing or Eliminating Online Obscenity Law as an Alternative to Contemporary Community Standards." *Communication Law and Policy* 9 (2004): 25.

Conrad, Barnaby. *Absinthe: History in a Bottle*. San Francisco: Chronicle Books, 1997.

Davis, Fred. "Deviance Disavowal: The Management of Strained Interaction by the Visibly Handicapped." Pp. 119-38 in *The Other Side: Perspectives on Deviance*, edited by Howard S. Becker. New York: Free Press of Glencoe, 1964.

Delgado, Richard. "Words that Wound: A Tort Action for the Racial Insults, Epithets, Name-Calling." Pp. 131-40 in *Critical Race Theory: the Cutting Edge*, Second Edition, edited by Richard Delgado and Jean Stefanic. Philadelphia, PA: Temple University Press, 1995 [1982].

Demac, Donna, and John Downing. "The Tug-of-War Over the First Amendment." Pp. 112-27 in *Questioning the Media: A Critical Introduction*," edited by John Downing, Ali Mohammadi, and Annabelle Sreberny-Mohammadi. Thousand Oaks, CA, London and New Delhi: Sage Publications, 1995.

Demac, Donna A., with Liching Sung. "New Communication Technologies and Deregulation." Pp. 277-92 in *Questioning the Media: A Critical Introduction*," edited by John Downing, Ali Mohammadi, and Annabelle Sreberny-Mohammadi. Thousand Oaks, CA, London and New Delhi: Sage Publications, 1995.

Durkheim, Emile. *The Division of Labor in Society*. Trans. W. D. Halls. New York: The Free Press, 1984 [1893].

———. *The Rules of the Sociological Method*. Trans. W. D. Halls. New York: The Free Press, 1982 [1895].

———. *Suicide*. Ed. George Simpson. Trans. John A. Spaulding and George Simpson. New York: The Free Press, 1951 [1897].

Earth and Fire Erowid. "Data Points in the Void." *Erowid Extracts: A Psychoactive Plants and Chemicals Newsletter*. Number 3, October 2002: 6-7.

Ellickson, Robert. *Order Without Law: How Neighbors Settle Disputes*. Cambridge, MA: Harvard University Press, 1991.

Erowid Extracts: A Psychoactive Plants and Chemicals Newsletter. Number 3, October 2002.

Etzioni, Amitai. *The Spirit of Community: Rights, Responsibilities, and the Communitarian Agenda*, 1992.

———. "Old Chestnuts and New Spurs." In *New Communitarian Thinking: Persons, Virtues, Institutions, and Communities*, edited by Amitai Etzioni, 16-36. Charlottesville and London: University Press of Virginia, 1995.

————. "A Matter of Balance, Rights and Responsibilities." In *The Essential Communitarian Reader*. Edited by Amitai Etzioni. Lanham, MD: Rowman and Littlefield Publishers, Inc., 1998, xi-xxiv.

————. "The Responsive Communitarian Platform: Rights and Responsibilities." In *The Essential Communitarian Reader*. Edited by Amitai Etzioni. Lanham, MD: Rowman and Littlefield Publishers, Inc., 1998: xxv-xxxix.

Finkelman, Paul. "Cultural Speech and Political Speech in Historical Perspective." *Boston University Law Review* 79 (1999): 717.

Fire Erowid. "Face to Face with NIDA: A Conference on Drugs, Youth, and the Internet." *Erowid Extracts: A Psychoactive Plants and Chemicals Newsletter*. Number 3, October 2002: 2.

Fire and Earth Erowid. "Surveying Erowid." *Erowid Extracts: A Psychoactive Plants and Chemicals Newsletter*. Number 3, October 2002: 8-9.

Fire and Earth Erowid, and Sophie. "Challenges of Visionary Parenting." *Erowid Extracts: A Psychoactive Plants and Chemicals Newsletter*. Number 3, October 2002: 10-11.

Froomkin, A. Michael. "The Internet as a Source of Regulatory Arbitrage." Pp. 129-63 in *Borders in Cyberspace: Information Policy and the Global Information Infrastructure*. Cambridge, MA and London: MIT Press, 1997.

Gandy, Oscar H., Jr. "Tracking the Audience: Personal Information and Privacy." Pp. 221-37 in *Questioning the Media: A Critical Introduction,"* edited by John Downing, Ali Mohammadi, and Annabelle Sreberny-Mohammadi. Thousand Oaks, CA, London and New Delhi: Sage Publications, 1995.

Gatson, Sarah N. "Illegal Behavior and Legal Speech: Discourse about Drug Use in an Internet Community/Network." Presented at Southwestern Social Science Association Meetings, San Antonio, TX, April 17, 2003.

————. "Youth, Drugs, Internet Discourse: A new arena of community and citizenship. *PISTA 2004 Proceedings*, Orlando: International Institute of Information and Systematics, 2004a.

————. "When Do Young People Become Community Members and/or Citizens?" Summer Institute on Digital Empowerment: The Internet and Democracy, Center for Digital Literacy, Syracuse University, Syracuse, NY, July 8-9 (via Telecast), 2004b.

Gatson, Sarah N., and Amanda Zweerink. "Choosing Community: Rejecting Anonymity in Cyberspace." *Research in Community Sociology*, vol. 10 (2000): 105-37.

————. "'Natives' Practicing and Inscribing Community: Ethnography Online." *Qualitative Research*, 4, no. 2 (2004a): 179-200.

————. *Interpersonal Culture on the Internet—Television, the Internet, and the Making of a Community*. Lewiston, NY: The Edwin Mellen Press: Studies in Sociology Series, no. 40, 2004b.

Gibson, Melissa. "Voluntary Use, Risk, and Online Drug use Discourse." in *Real Drugs in a Virtual World: Drug Discourse and Community Online*, edited by Edward Murguía, Melissa Tackett-Gibson, and Ann Lessem. Lanham, Md.: Lexington Books, 2007.

Giddens, Anthony. *The Constitution of Society*. Berkeley: University of California Press, 1984.

Godwin, Mike. *Cyber Rights: Defending Free Speech in the Digital Age*. July 1998.

Goldring, John. "Netting the Cybershark: Consumer Protection, Cyberspace, the Nation-State, and Democracy. Pp. 322-54 in *Borders in Cyberspace: Information Policy and the Global Information Infrastructure*. Cambridge, MA and London: MIT Press, 1997.

Goode, Erich. "Legalize it? A Bulletin from the War on Drugs." *Contexts*, Vol. 3, no. 3 (2004):19-25.

Gurak, Laura J. *Persuasion and Privacy in Cyberspace: The Online Protests over Lotus Marketplace and the Clipper Chip.* New Haven, CT: Yale University Press, 1997.

Hickok, jr., Eugene W., editor. *The Bill of Rights: Original Meaning and Current Understanding.* Charlottesville, VA and London: University of Virginia Press, 1991.

Kitsuse, John I. "Societal Reaction to Deviant Behavior: Problems of Theory and Method," Pp. 87-102 in *The Other Side: Perspectives on Deviance*, edited by Howard S. Becker. New York: Free Press of Glencoe, 1964.

Komasara, Tiffany. "Planting the Seeds of Hatred: Why Imminence Should No Longer be Required to Impose Liability on Internet Communications." *Capital University Law Review* 29 (2002): 835.

Krotoszynski, Ronald J. "Childproofing the Internet." *Brandeis Law Journal* 41 (2003): 447.

Lee, Orville. "Legal weapons for the weak? Democratizing the force of words in an uncivil society." *Law and Social Inquiry*, 26, no. 4 (2001): 847-92.

Lessig, Lawrence. *Code, and Other Laws of Cyberspace.* New York: Basic Books, 1999.

Lionnet, Francois. "Autoethnography: The An-Archic Style of *Dust Tracks on a Road.*" Pp. 164-95 in *The Bounds of Race: Perspectives in Hegemony and Resistance*, ed. D. LaCapra. Ithaca, NY and London: Cornell University Press, 1991.

Lipschultz, Jeremy Harris. *Free Expression in the Age of the Internet: Social and Legal Boundaries.* Boulder, CO: Westview Press, 2000.

Loader, Brian, and Douglas Thomas, editors. *Cybercrime: Law Enforcement, Security and Surveillance in the Information Age.* New York: Routledge, 2000.

MacKinnon, Richard. "The Social Construction of Rape in Virtual Reality." Pp. 147-172 in *Network and Netplay: Virtual Groups on the Internet*, edited by Fay Sudweeks, Margaret McLaughlin, and Sheizaf Rafeli. Menlo Park, CA, Cambridge, MA, and London: AAAI Press/MIT Press, 1998.

Marcus, George. *Ethnography Through Thick and Thin.* Princeton: Princeton University Press, 2000.

Mayer-Schonberger, Viktor and Teree E. Foster. "A Regulatory Web: Free Speech and the Global Information Infrastructure," Pp. 235-54 in *Borders in Cyberspace: Information Policy and the Global Information Infrastructure.* Cambridge, MA and London: MIT Press, 1997.

Malley, Wendi. "Blogging—A Cyber Portal: Changing Audiences to Publics." Typescript, 2003.

Marshall, Anna-Maria, and Scott Barclay. "In Their Own Words: How Ordinary People Construct the Legal World." *Law and Social Inquiry*, Vol. 28 (3): 617-28.

Merry, Sally Engle. *Getting Justice and Getting Even: Legal Consciousness Among Working-Class Americans.* Chicago: University of Chicago Press, 1990.

———. "Resistance and the Cultural Power of Law." *Law and Society Review*, Vol 29, No. 1 (1995): 11-26.

Merton, Robert K. "Social Structure and Anomie." *American Sociological Review*, Vol. 3, No. 5. (Oct., 1938), pp. 672-82.

Mitchell, William J. *e-topia: "Urban life Jim, but not as we know it."* Cambridge, MA and London: The MIT Press, 2000.

Morris, Aldon. *Origins of the Civil Rights Movement: Black Communities Organizing for Change.* New York: The Free Press, 1984.

Murguía, Ed, Sarah N. Gatson, and Melissa Tackett-Gibson. "Youth, Technology, and the Proliferation of Drug Use." Grant, National Institute on Drug Abuse/ National Institute of Health, 2001.

Nielsen, Laura Beth. "Situating Legal Consciousness: Experiences and Attitudes of Ordinary Citizens about Law and Street Harrassment." *Law and Society Review*, Vol. 34 (4): 1055-90.

Pinchbeck, Daniel. *Breaking Open the Head: A Psychedelic Journey into the Heart of Contemporary Shamanism.* New York: Broadway, 2003.

Rabban, David M. "Bad Tendency Test." Pp. 54-55 in *The Oxford Companion to the Supreme Court of the United States.* Oxford: Oxford University Press, 1992.

Redden, Jim. *Snitch Culture: How Citizens are Turned into the Eyes and Ears of the State.* Los Angeles: Feral House, 2000.

Reed, O. Lee. "The State is Strong but I am Weak: Why the Imminent Lawless Action Standard Should Not Apply to Targeted Speech that Threatens Individuals with Violence." *American Business Law Journal* 38 (2000): 177.

Reidenberg, Joel. "Governing Networks and Rule-Making in Cyberspace." Pp. 84-105 in *Borders in Cyberspace: Information Policy and the Global Information Infrastructure.* Cambridge, MA and London: MIT Press, 1997.

Rheingold, Howard. *The Virtual Community: Homesteading on the Electronic Frontier.* Reading, MA: Addison-Wesley Publishing Company, 1993.

Riley, Donna M. "Sex, Fear and Condescension on Campus: Cybercensorship at Carnegie-Mellon." Pp. 158-68 in *Wired Women: Gender and New Realities in Cyberspace.* Seattle, WA: Seal Press, 1996.

Rodman, Gilbert B. "The net effect: The public's fear and the Public Sphere." Pp. 9-48 in *Virtual Publics: Policy and community in an electronic age,* edited by Kolko, Beth E., (New York: Columbia University Press), 2003.

Rohall, David, editor. "From the polls: Illegal Drugs." *Contexts,* Vol. 3 (2): 2004: 60.

Ross, Susan Dente. "An Apologia to Radical Dissent and a Supreme Court Test to Protect It." *Communication Law and Policy* 7 (2002): 401.

Scheppele, Kim Lane. "Legal Theory and Social Theory." *Annual Review of Sociology,* Vol. 20. (1994): 383-406.

Schiller, Herbert I. "The Global Information Highway: Project for an Ungovernable World." Pp. 17-34 in *Resisting the Virtual Life: The Culture and Politics of Information,* edited by James Brook and Iain A. Boal. San Francisco: City Lights, 1995.

Sclove, Richard E. "Making Technology Democratic." Pp. 85-104 in *Resisting the Virtual Life: The Culture and Politics of Information,* edited by James Brook and Iain A. Boal. San Francisco: City Lights, 1995.

Smith, Christine B., Margaret L. McLaughlin, and Kerry K. Osbourne. "From Terminal Ineptitude to Virtual Sociopathy: How Conduct is Regulated on Usenet." Pp. 95-112 in *Network and Netplay: Virtual Groups on the Internet,* edited by Fay Sudweeks, Margaret McLaughlin, and Sheizaf Rafeli. Menlo Park, CA, Cambridge, MA, and London: AAAI Press/MIT Press, 1998.

Smith, Dorothy. *Texts, Facts, and Femininity: Exploring the Relations of Ruling.* London and New York: Routledge, 1990.

Sobel, David, editor. *Filters and Freedom: Free Speech Perspectives on Internet Content Controls.* Electronic Privacy Information Center, 1999.

Stembridge, Patricia J. "Adjusting Absolutism: First Amendment Protection for the Fringe." *Boston University Law Review* 80 (2000): 907.

Steinberg, Laurence, and Laura H. Carnell. "Testimony before the Subcommittee on Crime of the U.S. House Judiciary Committee." American Psychological Association Congressional Testimony, March 11, 1999. <http://www.apa.org/issues/pstein berg.html> (30 Apr. 2003).

Steinberg, Laurence. "Should Juvenile Offenders be Tried as Adults? A Developmental Perspective on Changing Legal Policies." January 19, 2000. <http://www.jcpr.org /authors_otherwork/steinbergpapers.html> (30 Apr. 2003).

Sumner, William Graham. *Folkways*. Boston: Ginn and Co., 1940 [1906].

Sutton, Laurel A. "Cocktails and Thumbtacks in the Old West: What Would Emily Post Say?" Pp. 169-187 in *Wired Women: Gender and New Realities in Cyberspace*." Seattle, WA: Seal Press, 1996.

Talmadge, Joe. "The Flamers Bible." 1987. <http://www.netfunny.com/rhf/jokes/88q1/ 13785.8.html>

Travers, Ann. *Writing the Public in Cyberspace: Redefining Inclusion on the Net*. Garland Studies in American Popular History and Culture, edited by Jerome Nadelhaft. New York and London: Garland Publishing, Inc, 2000.

Walzer, Michael. *Spheres of Justice: A Defense of Pluralism and Equality*. New York: Basic Books, 1983.

———. *What it Means to be an American*. New York: Marsilio Publishers Corp., 1992.

———. "The Communitarian Critique of Liberalism." In *New Communitarian Thinking*, edited by Amitai Etzioni, 53-70. Charlottesville and London: University Press of Virginia, 1995.

Wallace, Jonothan, and Mark Mangan. *Sex, Laws, and Cyberspace: Freedom and Censorship on the Frontiers of the Online Revolution*. New York: Henry Holt, 1997.

Zimmer, Lynn and John P. Morgan. *Marijuana Myths Marijuana Facts: A Review Of The Scientific Evidence*.

Zweerink, Amanda, and Sarah N. Gatson. "WWW.Buffy.Com: Cliques, Boundaries, and Hierarchies in an Internet Community." In *Fighting the Forces: What's at Stake in Buffy the Vampire Slayer*, edited by Rhonda Wilcox and David Lavery. Lanham, MD: Rowman and Littlefield, 2002.

Chapter Ten

Music as a Feature of the Online Discussion of Illegal Club Drugs

Joseph A. Kotarba

Musicians, critics, journalists, and fans have consistently noted the close relationship between popular music and illegal drug use. Some observers frame this relationship negatively, for example, as a path of self-destruction for otherwise talented artists and fans. David Crosby, a member of the rock band Crosby, Stills, and Nash, for example, wrote a very popular autobiography describing the accumulating and debilitating legal and artistic problems in his life attributed to twenty-five years of playing rock and roll while consuming abusive amounts of marijuana, heroin, and cocaine (Crosby and Gottlieb 1988). Other observers have focused on the positive aspects of this relationship, for example, the creative impact drug use has had on musical composition, performance and consumption. In their sympathetic biography of Jim Morrison, Jerry Hopkins and Danny Sugarman (1980:18) compared the late leader of the Doors to other, self-destructive yet creative, poets:

> The romantic notion of poetry was taking hold: the "Rimbaud legend," the predestined tragedy, were impressed on his consciousness; . . . the alcoholism of Baudelaire, Dylan Thomas, Brendan Behan; the madness and addiction of so many more in whom the pain married with the visions. . . . To be a poet entailed more than writing poems.

In either case, members of the world of popular music have developed a culture that has provided a rich, functional, and complex discursive tradition through which they talk, disseminate information, and share experiences about this relationship. The nature of this discourse has taken different forms over the years as styles of illegal drug use, the objectives of drug talk, and the media by which this talk takes place have all evolved. I refer to this mode of communication as *drugmusictalk*.

The purpose of this report is to describe the structure and function of illegal drug talk involving music that takes place on the Internet. The case in question is *drugmusictalk* that takes place on one particular Internet site dedicated to education for and discussion of responsible drug use, especially designer drugs (e.g., MDMA). Thus, the *drugmusictalk* that takes place on this site provides us with a glimpse into the world of club or designer drugs, rave parties, dance, and high technology. As we will show, music is a key feature of this world.

I will begin with a quick summary of the history of the relationship of popular music and illegal drug use, followed by a brief discussion of the evolution of *drugmusictalk*. I will then describe the larger project from which the present analysis is derived. I will then present an illustrated inventory of the ways music enters interaction on the Internet.

The Relationship of Popular Music and Illegal Drug Use

Lay and professional observers alike have noted the parallel emergence of popular music styles and styles of illegal drug use since the early twentieth century.[1] This is not to say that all musicians and their audience members use drugs. The point is that these two behaviors combine into lifestyles that, at various times, are seen to be morally, legally, aesthetically, and/or medically problematic.

Blues music is one of the earliest styles of distinctively American popular music. The blues emerged in the rural south and river towns in the 1910s and '20s (Jones 1963). Although not the exclusive drug of choice among blues performers and their fans, marijuana has been central to the blues scene.

Jazz, on the other hand, emerged at more or less the same time as the blues, but in different locations (e.g., New Orleans and New York) and in a different style of scene (i.e., middle-class African Americans and Caucasians in a very urban setting). Heroin has traditionally been the drug of choice in the jazz world (Jones 1963: 201-2). Jack Kerouak, Allen Ginzburg and other Beat Poets of the 1950s and early '60s emulated Charley Parker, John Coltraine and other jazz artists' hard lifestyles as well as their work (e.g., Kerouak 1957).

As we move into the 1960s, the presence of recreational drugs became widespread among white adolescents creating and appreciating rock music. Marijuana, alcohol, methamphetamines, and LSD were all very popular, as were the Rolling Stones, Led Zepplin, and the Beatles. As Lenson (1998) notes, drug use was central to the rock music scene of the late 1960s because it enhanced the experience of improvisation—the hallmark of countercultural music—for both artist and audience.

The 1970s witnessed a decline in the countercultural power of rock music, within the context of an increasingly conservative, post-Vietnam war society. Thus, we witness the emergence of disco music. The French template for the high-level nightclub arrived almost intact in New York in the early 1960s and became an enactment zone for the drug-fueled cultural revolution of that era. Renamed disco in the 1970s, the phenomenon confronted white, heterosexual America with the newly liberated gay lifestyle, which was dedicated to frantic promiscuity and avid, recreational drug use (Braunstein 1999). Popular drugs in the disco scene included amphetamines, methamphetamines, and marijuana.

The 1970s and 80s marked the advent of punk music and heavy metal. To some degree, both styles of music involved a return to injected drugs among marginal segments of their respective scenes. Pills of various kinds were popular as well as traditional marijuana. The wholesale movement of illegal drug used out of the black community and into white suburbia. Punk and heavy metal elicited great response from agents of social control—for example, Tipper Gore's Parents Music Resource Center—saw the link between these styles of music and drug use as a major social problem (Nostalgie de la boue 1994).

The emergence of rap music as a feature of the broader hip hop culture in the late 1980s was marked by accusations of its links with growing crack cocaine use (McKinney 1989). Freebasing became popular in the 1980s among the poor because it was an economical way for them to enjoy cocaine—traditionally a vice for the privileged classes. Beginning with Grandmaster Flash and the Furious Five's "White Lines," crack cocaine became a central theme in rap music (George 1998).

The popular music scene at the turn of the century witnesses a major turn to techno music, dance, and designer drugs. Specifically, the literature on techno dance and drugs gives one the impression that the increase in the use of ecstasy (MDMA) and the growth of the Techno-scene, especially in the 1990s, are highly correlated. Hitzler (2002) notes that most ecstasy users identify themselves as fans of techno/rave, and those techno fans who consume drugs generally prefer ecstasy.

In summary, drugs and popular music go together. Fans not only emulate musicians' work, but also emulate their lifestyles. Fans—and critics—see drugs as central to the music, creativity, and overall social world of the musician.

Music and Drug Discourse

The popular music scenes mentioned above have all contained media for communicating about music and drugs. This communication generally reflects current musical fashion, drug fashion, and available as well as desirable communications media. Traditionally, members of the scene are likely to engage in *drugmusictalk* for the following purposes (among others):

To share aesthetic drug-music experiences;

To discuss fit between particular drugs and styles of music;
To share affective aspects (e.g., sexual) of drug-music experiences;
To discuss economic and/or ideological aspects of drug-music experiences; and
To discuss personal tastes in music.

Blues music emerged as a true form of folk music. The role distinction between performers and audience members was often blurred, as the early acoustic versions told stories relevant to the everyday lives of all present. Repression from slavery at first and sheer racism later resulted in the need for blues musicians and their audiences to disguise the potentially radical messages in the music through the heavy use of vernacular, metaphor, and innuendo. Much of the *drugmusictalk* was encoded in the lyrics of the blues singers' songs. In addition, the lack of financial resources among blues people precluded any attempt at developing an organized press to disseminate information about the blues (Jones 1963).

The jazz community, oriented around a largely instrumental style of music, developed *drugmusictalk* in the actual conversation between and among musicians, audience members, and others. The jazz world's version of *drugmusictalk* was created by both the heroin addict and the musician together. As Leroy Jones observes:

> much of the "hip talk" comes directly from the addicts jargon as well as from the musician's. The "secret" bopper's and (later) hipster's language was the essential part of a cult of redefinition, in terms closest to the initiated (1963:202).

Since jazz became a widely recorded style of music, marketed in entertainment centers like New York City, it produced a new form of *drugmusictalk* known as jazz music criticism. Newspaper writers and magazines dedicated to jazz (e.g., Down Beat) conveyed the jazz vernacular to a wide and appreciative audience.

The development of rock music in the late 1960s was marked very clearly by an appropriation of African-American music and drug behavior by white adolescents (Frith: 1983:192). *Drugmusictalk* was imbedded in lyrics of many songs (e.g., "White Rabbit" by the Jefferson Airplane among many others). As Market (2001) observes:

> Songs dealing with illegal drugs have long dotted popular music. It was not until the aftermath of the sixties youth counterculture, however, that drug lyrics became a recurring musical motif. In the decades since, the lyrical treatment of drugs has undergone change. Heroin and cocaine have largely, though not exclusively, been treated antagonistically, with the animosity toward cocaine becoming more pronounced after crack cocaine was introduced in the mid-1980s. Marijuana, on the other hand, has generally been perceived as innocuous, if not positively assessed, and this treatment has crossed the decades into the nineties. In more recent years, however, the positive assessment of marijuana has undergone change, with younger musicians more likely to decry the harm that drugs

do than older musicians do. This prosocial aspect of contemporary popular music has been largely ignored.

Nevertheless, the counterculture was also notable for developing its own media, in the form of *Rolling Stone Magazine,* FM radio, concert fliers, and community-based, underground newspapers and journals (Market 2001). Later versions of rock music, specifically heavy metal, developed a multi-media and diverse system of *drugmusictalk.* Pareles (1988) uses the early career of popular heavy metal band Metallica to illustrate this point:

> The heavy metal rock band Metallica plays loud, high-speed music with lyrics that dwell on dark subjects such as death, madness, nuclear war, and drug abuse. While adhering to heavy metal's basic tenets, the members of Metallica rebel against many of the conventions associated with hard rock music and refuse to package themselves for mass consumption. The band has never made a video for MTV, and, until the advent of all-hard-rock radio formats, Metallica albums were never played on commercial rock radio stations. Nonetheless, the group has attracted an avid following, mainly through tours, heavy metal fan publications, some college radio exposure, and word of mouth. With its skittish, hard-driving music, Metallica manages to avoid the formulaic quality of most heavy metal bands.

Interestingly, agencies of social control, such as the American Medical Association, interpret the lyrics in heavy metal music as dangerous to its fans:

> The American Medical Association (AMA) and the American Academy of Pediatrics have voiced concerns about certain lyrics used in heavy metal and rap music. The AMA says that messages in these genres may pose a threat to the physical health and emotional well being of particularly vulnerable children and adolescents. The AMA has identified six potentially dangerous music themes: drug and alcohol abuse, suicide, violence, satanic worship, sexual exploitation, and racism . . . . the AMA and the Academy of Pediatrics support voluntary regulation and increased social responsibility in the music industry (Levine 1991).

The next, major development in *drugmusictalk* was the vernacular that emerged from the rap and hip hop culture. Recall that the *drugmusictalk* that accompanied the blues and jazz was intended to keep the larger, often judgmental, white world out of the scene. Rap and hip hop talk, on the other hand, shoved the everyday life world of Compton, CA, Brooklyn, NY, and Houston, TX right in the face of anyone willing to listen—whites included.

The most recent style of *drugmusictalk* is the communications taking place in the world of rave parties, techno dance clubs, and designer drugs—the focus of the present study. Collin (1997) notes how communications in the early days of rave/techno was fairly primitive—artistic party fliers, dance floor talk, and telephone. Cummings (1994) writes about the eventual development of fan magazines dedicated to rave/techno, and the most dramatic development of all:

the widespread use of the Internet among rave/techno participants.

Research Question and Theoretical Framework

The primary research question for this study is: What is the structure and function of *drugmusictalk* in the context of the Internet? The theoretical framework for this study is Symbolic Interaction. In general, symbolic interactionism argues that social life is constructed through various forms of interaction. People interact with each other to pragmatically arrive at solutions for shared problems, most often experienced in concrete situations. As Altheide (2001) notes, the logic and format the shapes the Internet largely determine the structure and function of discourse that occurs on the Internet. As Denzin (1997) argues, contemporary media, especially the Internet, empower speakers to create realities. Media is more than just the message: it is the world in which lives are lived, cultures are constructed. As Kotarba (2002) indicates, the Internet specifically allows people to recreate the self situationally and constantly.

In summary, symbolic interactionism views the Internet as if it were any other setting studied by sociologists. The medium has evolved and the vernacular as changed, but we can expect many of the same pressing, everyday issues to be at the heart of electronic communication.

Method of Inquiry

There are two components to this study. First, I conducted a content analysis of the Internet site's (a pseudonym) forums. The following are the forums in which music and drug topics are discussed:

> Music and DJs
> Words
> Drug Culture
> Good Experiences
> Bad Experiences

I searched for elements of *drugmusictalk* and categorized them inductively, following the logic of grounded theory. Grounded theory is a style of social scientific analysis that operates with the logic of induction. Research begins with observations of activities. Patterns emerge from a constant comparison of cases continuously chosen to seek variation. When variation is exhausted, findings are organized into categories, which themselves are exhausted (Charmaz 2002). In the present study, the essential research question is: What are the various ways participants in the Internet site discuss music and drugs?

Second, I analyzed the responses to three items included in the Texas A&M University Survey 2002 on club drugs on the Internet. (Please see the detailed description of the survey above.)

Traditional *Drugmusictalk* Themes

The participants in the Internet site talk about the five traditional *drugmusictalk* topics listed above. I will present examples of each theme. The reader should keep in mind that these themes typically overlap in actual posts. Any one post can contain more than one theme.

Sharing aesthetic drug-music experiences
The Internet site participants talk about their drug music experiences as if they were like artistic or theatrical performances. This discourse is frequently sprinkled with efforts to describe the sublime pleasures associated with dancing. The stories tend to be long, well-written narratives. For example:

> Alright, for some background information . . . I have taken a fair amount of pills in my pill career, the most taken in a night being 3 and the most taken at one time being 1. I have never double dropped because I'm too worried about the tolerance that it might develope for future rolls. I usually have a relatively high tolerance to mdma, one pill not taking me as high as most but I don't usually chase highs. I prefer to take pills to enhance the music and the dancing when I go out. I usually start off on a whole then take halves to keep the peak up until I start the downward spiral.
> Other drugs I've had are weed, speed, rush (amyl), acid and dxm.
> The pills I had were white MX's, very clean strong mdma if you haven't heard of them.
>
> **11pm**—Arrived at a one of my preferred nightclubs for an uplifting trance night with a few friends and was gobsmacked by the production of the club-night, starting off an awesome night. Sat around and started to feel the music and dance in my seat, already mystified by the lasers on the dancefloor.
>
> **11:30pm**—Shafted my first pill. It had to be one of the most uncomfortable things I have ever done to myself and the only thing that pushed me knuckle past knuckle was the thought of a pill high like never before. Apart from the sensation of a finger briefly being up my exit hole, it felt okay and bearable. If anything there was a wierd feeling like in 20 minutes I would need to go to the toilet but I knew it was just the pill dissolving. I sat back down and got into the mood of the night. One friend had shafted a whole and another had shafted a half and swallowed the other. I knew what to expect of these pills so I played the waiting game, eager to see what an 80-90% absorbtion rate from different mdma receptors would be like.
>
> **12:00am**—One of my favourite DJs was going to start in half and hour and the dance floor hadn't fully packed out yet so we went up to dance. As soon as I stood up I though "oooooh I feel good!" There was a feeling of warmth going all over my body but it felt like it was coming from my lower abdomen or even

inside my ass! Probably quite true. Starting dancing and getting into the music forgetting about the pill but already feeling great.

12:15am—Huge warm pill rushes filled my whole body, from the tips of my toes to the top of my head. They felt like an orgasm taking over my whole body but without the electrified feeling that real orgasms have (if you can imagine that). The feeling seemed most intense in the lower body again. Both other friends who had shafted had huge grins and could barely dance.

12:30am—Oh the bliss! I was so overwhelmed by these warm rushes of pleasure all over my body that I couldn't even talk! Dancing was cut down to the basics. A little bit of a bop and jelly like arm movements were all I could manage, a huge ear to ear grin filling my face. From this point on you couldn't say that I was off my trolley, the trolley had knocked me over! Both friends looked like they were having a time as good as mine but none of us had to say a word because we all understood what each of us were feeling.

1:00am—The rushes had settled down and I was into an extremely intense peak. The lasers and the music had taken my complete attention and I honestly couldn't tell you if I were dancing or not. I probably was, just like I was off my head, haha! I had forgotten all about my favourite DJ being on but the music was awesome uplifting trance, a perfect pillow for my state. The sweet melodies were bringing extreme emotions of happyness, nostalgia, sorrow and excitement from deep within me. An ear to ear smile being the only expression of it. I closed my eyes and danced like I didn't care.

2:30am—I had danced like an animal the whole time! Full of energy, feeling the music like it was a part of me or a reflection of things in my life. One of my favourite local trance acts had gone and finished and could barely remember it but I knew it was great. One of my other shafting buddies was going to have another half a pill and offered me the other half. How could I refuse? Entering the toilet cubicle cue I started chatting with a random gurner, the conversation starting off with him coming to me with "I'm glad it looks like someones having a night as good as I am." "God I don't look that fucked do I?" I thought to myself but had a good conversation with a birthday gurner. My mate came out of the cubicle and gave me half his pill and I went into the cubicle. A quick decision came over me to shaft this half as well. I thought "I'm so high right now that I could probably get lost in this cubicle but how high can I get?" Down periscope, another half up the poo whole. An instant rush hitting me as I did it, probably just a placebo effect. Back to the dancefloor.

3:00am—Well and truly munted. Pill fucked in its truest sense. 5 minutes would pass and feel like an hour. My mate that shared his pill with me is a really good friend to go out with because we always seem on the same wavelength with our drugs and moods. In the middle of dancing we both engaged in a pill fucked communication where we would both mime out musical instruments. Everything from drums, guitar, double bass, flute and even the triangle! No instrument was spared! Both of us completely involved in it, completely forgetting about the music and the rest of the dancefloor around us. When we both thought there we no instruments left we had a huge laugh and went back to dancing.

3:30am—An intense peak putting me right into the clouds. Gurning and hallu-
cinating. What more could you ask for? I had never actually gurned before, I
had been missing out until this point. Slight OEVs such as a water bottle on the
floor looking like a mobile phone with multiple flashing lights as buttons, the
back wall of the club behind the DJs was black with little star lights to make it
look like the nights sky and I was convinced that it was at one stage. Plane's
guidance lights slowly flashing their way across the sky. The CEVs were quite
strong and vivid as well. Callidascope (spelling?) images morphing and changing colours behind my eyelids, so beautiful! A person on the dancefloor tried to
start conversation with me but I was too fucked to engage in it. Usually I love
talking to randoms at clubs but this time I was in so much bliss at that time that
I didn't need to. The water in my bottle felt like thick cream as I swallowed it,
almost tasting sweet. I got lost in the music and danced the night away.

6:00am—EEEEEK! The lights!!!! The lights in the club had gone on already,
time to go home. Where had the last couple of hours gone!!?!?! It felt like I had
only danced to a few songs and the night was over already. I knew I had a great
time and had some faint memories of leaving the club and going for a mission
into the park across the road to have a rest. While we were walking around in
the park I could remember feeling so floaty that I couldn't feel my body and
was not conscious that I was walking. It was just my smiley face head floating
around 8 feet above the ground.
We got outside the club and without the music and club atmosphere I had already begun to come down. No more peak, no more small plateu, just a scattered comedown before me. I didn't feel grumpy at all, at this stage, just the
feeling of quiet satisfaction you get as you start to come down off a good pill.
We all sat down at a coffee shop watching people go to work in their business
suits and uniforms feeling extremely seedy. No decisions could be made. We
just sat there until someone would do something constructive, my comedown
getting quite bad and grumpy. We decided to take a ferry to the beach and get
some Robitussin DX. Thats another story, the trip report is around here somewhere.
The whole next week I was grumpy and depressed. Completely and utterly
robbed of all my seratonin and it felt like my brain wasn't in any hurry to re-
build it. If you only absorb around 40% of a pill when you swallow it and 80-
90% when you shaft it, I had just double dropped then topped up with a whole
a couple of hours later. That night caused me to develope a tolerance to pills
but I'm not complaining at all. That night sent me as high I could have liked to
be. The whole body sensation was like nothing I could describe, the only thing
I knew was that it was overwhealmingly good!
Will I ever do it again? I doubt it unless I want to remember whats its like to be
a gurner again.
I hope you got something from my report ☺Thanks for reading if you got this
far.)

Music plays a key role in this type of adventure. It provides something to do, as
well as a context for other things to do. Music also functions as something to

talk about, as well as something to use to display one's skill, at evaluating and critiquing key features of the techno and dance lifestyle, such as drugs.

Discussing the fit between particular drugs and styles of music
Internet site participants talk about the fit between drugs and music much like a chef designs a menu. The following postings illustrate the holistic nature of designer drug experiences:

> as everyone knows music is awsome as hell on E . . . but what music should i download to listen to when im rollin? can neone give me some songs?

And the following answers

> listen up. it all depends on wut u like when you are sober. Xtc will make it sound 10X better. do a search for what kidn of music do you like to listen to when rolling. I have a funny feeling this is gunna get shutdown or booted off to music forum.

And

> love to start off listening to a little rap when i am comin up, then when i start peaking i love to listen to techno . . . . try somethin like darude—sandstorm for some old school stuff, or maybe if u like the radio stuff, dj sammy we're in heaven (but i dont recommend it) i looooove to listen to zombie nation—oh oh oh oh oh . . . and on the comedown some pink floyd suits me, even though it is kinda depressing. so thats me, i dono about u.

And

> that isnt techno, its electronica, but more specificly cheese-trance and i probably wouldn't even call that commercial shit electronica.
> for me, trance/hard-trance is fkn awesome while peaking coz trance takes you on a journey 😊but it might not be for u, but id recommend giving it a try.
> try these:
> Plastic Boy—Live Another Life (Original Mix)
> Rank 1—Sensation 2003 Anthem
> nu nrg—butterfly
> Angelic—Can't Keep Me Silent (Dumonde Mix)
> BK—Revolution
> Cj Bolland—The Prophet (Original Mix)
> dave 202 and phil green—legends (club mix)
> DJ Air—Alone With Me (Flutlicht Mix)
> DJ Wag—The Darkness
> Green Court—Silent Heart (Flutlicht Remix)
> Kamaya Painters—Far From Over
> Haak—Frenzy
> frank trax—nebuchan (organ remix)
> matanka—lost in a dream (push remix)

Push—Till We Meet Again
Signum—Coming on Strong (SHOKK remix)
Andy B Jones—In Motion (Arrowhead Mix)
Dave 2002—The Klammt (Evacuation Remix)
The Mystery—Devotion (Tatana Remix)
NRC—Here Comes The Rain (Push Remix)
Shane 54—Vampire
DJ Tatana—Moments (flutlicht remix)
Signum—Cosmos
The Gift—Love Angel (M.I.K.E. mix)
Carlo Testi and Genjen—Contact (Junatik Pentagon Mix)
Sworn—Treatment Declined
Armin van Buuren—Precious
nrc—here comes the rain [push instrumental]
Dave Gahan—Dirty Sticky Floors (Junkie XL Dub Remix)
Cupa—Blaze
Origene—Sanctuary (Traveller Remix)
Delerium—After All (Satoshi Tomiie Remix)
Vivian Green—Emotional Rollercoaster (Above and Beyond Remix)
kate bush vs infusion—running up that hill
Mach 747—Invading Privately (29 Palms remix)
Motorcyle—As the rush comes
deepsky—talk like a stranger (sunday club mix)
Telepopmusik—Breathe (Markus Schulz Remix)
Jewel—Intuition (Markus Schulz Coldharbour Remix)
Jewel—Intuition (Attention Deficit Hi-Tek Mix)
pvd vs second sun—crush
Sander Kleinenberg—Buenos Aires
riva—time is the healer (armin mix)
23 Andain—Summer Calling (Airwave Club Mix)
Misja helsloot—First second / (signum remix)
DJ Energy—Arya
Walt—Wanna Fuck (original mix)
all awesome songs 😎

And

for rolling and dancing: hard house, progressive and techno
for rolling, chilling and enjoying the music: trance or psy trance because it
makes me trip in a beautiful way. hip hop suits me a lot better when stoned,
when rolling I need stuff on 120-140 BPM, but electronica in general (stuff like
Orbital or FSOL is amazing on E))

Sharing affective aspects (e.g., sexual) of drug-music experiences
The sexual dimension of drug-music experiences is an important topic for Internet site discussions. It perhaps illustrates best the social aspects of illegal drug use, as well as its impact on intimate relationships. For example,

so here's my story, it's a bit long

I'm going through an incredible time in my life, something which happens once in a lifetime.

I've done E 2 weeks ago for the last time, it was incredible. I've been doing E for a year now, once every 3 weeks and it's always been great. I'll always remember what happened the first time, it was a life changing experience. I'll always remember my last time because it was an experience which truly and fully openned my mind to some of the ways of friendship and love I haven't even imagined. I did it with other three people in a Summer Solstice Party in a forest. The after party was at a friend's house. My best friend, and now I know this. And that night was his first pill!!!

I'm in love with his ex-girlfriend, his most important girlfriend and it seems that she is in love with me. The last time on E he realized that, and he also realized that our love is real, this thing was something that strenghtened our friendship in the purest way. I still can't believe that he wants us to be together. Me and her, and me and him. And this morning I realized he even helped the whole thing. Man my last time on E was an incredible journey, maybe I'll do a trip report when I calm down and analyze fully the whole situation. Last friday I was at a club with two girls, really close friends, great house tunes, great place, great party . . . I had a pill in my pocket . . . and I didn't eat it!!! I was willing to eat it, a lot. But I didn't. I danced 4 hours in a row just on 2 drinks, it wasn't near as good as on E, but I was happy, so happy that I didn't eat the damn pill, and I was happy in general. The day after I was with her and some other people. she told me a lot of very important things, including she wasn't ever trying ecstasy, she wanted me to tell her everything about E, but she has decided that she won't do it. She also told me that she didn't mind if I took it in moderation. And she made me understand that she really believes in our relationship which is just starting. I'm 29 but feel like if I was 16, this is the most beautiful thing that has ever happened in my whole life. The night after our conversation, I didn't sleep, I had this pure euphoric feeling, well you know, way beyond what ecstasy could give you, and I even put some tunes on and it was incredible the way I was hearing the music, better than on E. Tonight I'm taking her for our first real date. And I'm thinking on quitting E. She will even let me go to Ibiza this summer, but I don't see the point anymore. My next pill is due two weeks from now, a big party, the biggest summer party in my town, organized by friends of mine. I'll be taking E for the last time, if I do. I won't be trying convincing her in trying E. She knows what she wants really well, and I think she just wants me, sober and no chemicals. maybe I won't be reading this board for a while. I love you people. I love E, it really made me the way I am now, and I feel like I'm on top of the world. But I know my next pill won't be that special thing, because there's something alive and real and it's so very special for me right now
Happy rolls and peace and love everybody.

Discussing economic and/or ideological aspects of drug-music experiences
Economic and/or ideological aspects of drug-music experiences are very common. Only certain participants, largely male and a bit older than most, pursue these topics. For example:

just because you download EDM doesnt mean that you are going to be safe and the RIAA wont touch you. be smart with your filesharing, if you are worried about being caught then do not share files . . . as soon as you download them move them to another folder that isnt shared.

also there is a program out there called peerguard (something like that). from what i understand it uses a list of known RIAA ip addresses and blocks them from connecting to your computer. again, this increases the changes that the RIAA wont catch you, but it does not make you 100% safe. the only way to really be 100% is to stop using p2p/file sharing apps all togeather, but really now, whos gonna do that?

think about it thou. how many ppl share files over the net? millions? more? how many people get caught? 1000? less?

And

yeah i wanna get some decks, a decent mixer with a beat counter(yeah i know im a shit cunt) and some headphones.
im in australia and i was wondering if this equip is ok
—2x Stanton STR830 direct drive turntables.
—1x behringer VMX200 mixer
—unsure about headphones yet.
this will set me back $1187 for the decks and mixer.
what i was wanting to know is this equipment any good for a beginner (i can mix alright) or should i just find something else.
this stuff is brand new.
Dfi☺

Discussing personal tastes in music
Participants talk about their favorite styles of rave music. The many terms used to identify musical styles allow respondents to custom design online identities. The many terms used to identify music styles also index the fragmentation of popular music culture in the postmodern era (Jameson 1991). The following is a partial list of rave music styles:

DnB
Progressive
Hard Trance
Industrial
Metal
Rock
Classical
Andrew's Sisters (40's group)
80's Love Songs
African tribal music
trance (deeeeeep!)
Rock
Classical
electronica
African tribal music

Drum and Bass
Jungle
UK Garage
Hip Hop
Hard Trance
NRG
Jungle
Breakbeat
Techno
Acid
Jazz
Alternative
pop
downtempo
breaks
techno
dnb
RnB
Reggae
Soul
Small elements of rock, certain tunes that I happen to like . . . Jam Music
Reggae
Jazz
UK Garage
Techno
Old Skool Hip Hop
Deep House
Classic Rock
Ska

An Internet site participant recently raised the following question in the "Music and DJs" forum: "What are your musical preferences?" The variety of sub-genres within the dance music scene is overwhelming. Typical answers included"

> "it depends on the day and my mood or my mission for the day."
> "i have such a broad range of musical preferences."
> hardcore punk
> jungle
> progressive house
> hard house
> ambient
> downtempo
> jam bands
> jazz
> classical
> other
> "how could i forget hiphop . . . see i love so much . . . good music is what it is"
> "i can listen to something different at any given moment . . . it just depends on

the moment."
"i do tend to get in ruts sometimes, but they usually don't last longer than a
week straight. It mainly happens when i hear something new that i fall in love
with, or when my jungle addiction takes over . . . I have to cut myself off after
about a week straight."

Interestingly, music fans who are not participants in the techno/rave/dance scene
have great difficulty differentiating among these various styles or subgenres.[2]
There are three possible explanations for this complex phenomenon. First, multiple musical styles represent a cultural value in the scene. High status is attached to members who are sophisticated in their taste in music. The ability to
differentiate and choose among many subgenres illustrates this sophistication.
Second, dance music by itself is pretty simple music. It is rhythm-intensive and
repetitive. Dance music lacks the textual complexity of lyrics. Participants in the
dance music scene may maintain if not rejuvenate interest in the scene by creating subgenres. Third, and perhaps most relevant to this study, outsiders are
oblivious to the nuances and variations within dance music because the nuances
and variations are only perceivable when high on drugs.

The Internet Allows People to Create and Perform
Situational Identities

As postmodern existentialism tells us, the contemporary experience of self is
complex and situational. We can be different things—to ourselves and to others—as the situation requires and our past experience suggests (Kotarba 2002b).
The Internet provides a great opportunity to create, develop, and project a self in
a way never available to older media. The Internet allows for long and immediate, narrative presentations of self. Describing one's self is not constrained by
the grammatical and stylistic constraints posed by older media. The great complexity of the site-structure of the Internet—with its chat rooms, forums, and
homepages—allows a wide variety of venues to write for a very specific and
directed audience, and audience as close to one's interests and expertise as ever
possible in the history of communications.

Yet, in our mass mediated, postmodern world, people can be different selfs
at different time in different situations. Internet site participants may be different
self-identities in other aspects of their lives, in front of other audiences.

I posted the following question on the "music and drugs" board: "Why do
so many people hate country music?" The following are typical responses:

"Good gawd!"
"How the f*ck did we get on the topic of country music!? I mean seriously,
MAN!!! What's going on!? This is the 21st century! Nobody loses their girl-
friend, their dog and their trailer anymore."
"They should update their shit—how about: Blue Screen Of Death Blues"
"Dude, just cause you hate contry doesnt mean you gotta knock it to peeps that

do. Just say you dont like it . . . . without the 'BARF' and shit like that. If your
a DJ then you should know the appreciation of music. I dont like country either
. . . but i know plenty of peeps that do . . . there is nothing wrong with that. I
cant say that i like being in a room of it playing for hours on end . . . but its def
not the end of the world. just like someone you know maynot like your music .
. . but might just put up with it."

Internet site participants locate their selfs directly in their drug experiences. Music makes these selfs all the more complex, elegant, desirable, esteemed, and
worthwhile. For example:

> So the quesion is When you are rolling what gets you in that "ecstasy" state
> more: hard pounding energetic music or smoother and gentler music?
> Personally for me its gentler music because when I'm rolling my mind can't
> really keep up with all the hard pounding intriquet sounds . . . it becomes too
> chaotic that I get annoyed . . . but when I listen to some smooth gentle trance
> that progresses very nicely from an ambient beginning to a climatic end I drift
> away into bliss. Music that is constantly energetic is too much for me, but I *can*
> handle music that is very energetic but in waves. And I can't stand extremely
> slow music, like I said before, song has to come in waves of energy or I get
> bored with it . . . I'm VERY picky while rolling. Oddly enough it is the oppo-
> site when I'm sober . . . I love hard pounding insane music that never ceases
> with it's energy instead of slow builds and dreamy soundscapes. Weird huh?
> What about you? Hard and Banging or Gentle and Progressive?

And

> depends on my point in the roll really, if im coming up or down, something
> gentle (preferably with some amazing vocals), but if im in the strong part of the
> roll i prefer some BANGIN hard music so i can dance my ass off)

The Connoisseurs Speak

One of the more interesting findings from survey data items on music and drugs
was the style by which Internet site participants answered the following question: "What types of music seem to go best with specific types of drug experiences?" There are three types of responses. The first type of response is simple
and direct. It mirrors the way earlier blues, jazz and rock audiences largely perceived their drug and music options:

> Classical and Rock are by far the best to listen to when high/drunk.

And

> Techno is best with ecstasy.

The second type of response illustrates the essential complexity of the techno/dance culture that lies beyond mere dichotomous choices:

> Smoking cannabis alone at home: any music I like also otherwise
> Smoking cannabis with friends at home, in a park or other controllable environment: electronic music with a rich sound environment (Astral Projection, Hallucinogen)
> Mushrooms: something peaceful and rich in sounds, as Pink Floyd or Twin Peaks soundtrack
> Ecstsasy: fast techno music

And

> Although individual preference plays a large part. I think fairly fast slightly hypnotic yet melodic music with anthemic breaks goes down well with most people on most drugs. Generally speaking trance, house, happy hardcore.

The third type of response illustrates a sophisticated mastery of the phenomenon:

> Trance/progressive/techno (not happy hardcore I may add)/ hardhouse—pills and cannabis. Trance makes you rush like mad, dancing to progressive stuff on pills is wicked and techno just makes a weird underground atmosphere. Hard house is good if you just want to get more off it, bit faster more energetic.
> Psychedelic Trance—obviously Lsd, shrooms e. t. c
> All well produced dance music sounds good when stoned.

And

> That is an entirely personal question (greatly varying between people), but for myself I would say that ambient techno (some orbital, Groove Armada) "chick music" (sarah mclachlan, tori amos, etc) are good for home experiences with one other friend. This makes the experience very calming, less chaotic and quite personal. Industrial techno, jungle, house and happy hardcore techno are lovely for a large gathering of people, say a party or rave. This keeps people active, moving and fluid.

And

> Music parallels the attitude of those drawn to it, and thus the type of feeling desired from the drugs. These are mass generalizations, but from my experience . . . . MDxx—Trance (the uplifting, almost spine tingling nature of the music parallels that of the feeling you get from ecstasy) Amphetamines—Hardcore or Drum n bass (the sped up basslines parallel the sped up nature of the drug) Marijuana—Downtempo (the chilled out nature of the music parallels the chilled out nature of the drug) Alcohol—House (im not quite sure why this works. it just does. Who doesnt like drinking and dancing all night to some disco? i suppose the relaxed funky attitude of the music parallels that of the drug)

The most sophisticated responses seem to come from older (twenties and thirties), college-educated, and male participants. They talk about music and drugs in ways very much like older wine connoisseurs talk about which cut of beef goes best with a frisky merlot. Interestingly, most respondents who go into great detail on this item list alcohol and place it very specifically in their inventories. Simpler responses tend to state that alcohol—or marijuana for that matter—go well with all sorts of music.

Conclusion

The Internet site appears to be a desirable and useful venue for talking about one's music, dance, and drug experiences holistically, as they occur in everyday life. The strength of Internet site, from the participants' perspectives, is that is provides the opportunity to these experiences with others very much like them. The liability, from a social control perspective, is that Internet site provides the opportunity to celebrate these experiences.

Drugmusictalk describes many different ways music relates to the designer drug experience for Internet site participants. Music is used to generate a meaning context for drug experience. Music functions as a stimulant for drug experiences. Listening to music provides ways to pass time during drug experiences, and to establish mood during drug experience. Music also has a distinctive social function in a world in which drugs, dancing, and other behaviors are very individualistic: it provides something to do together during drug experiences. Finally, music can serve as a depressant for drug experience, to counteract the high associated with rave/techno/designer drugs.

Notes

1. Obviously, the relationship of drug use (illegal or otherwise) to musical experiences can be traced much further into the past than the twentieth century. We limit our discussion, however, to popular music as a distinct feature of modern capitalistic culture. Popular music refers to music that is formulaic, mechanically or electronically reproduced, mass marketed, and disposable (Lull 1992).

2. I derive this observation from my ongoing research on popular music in everyday social life (e.g., Kotarba 2002a).

References

Charmaz, Kathy. "Grounded Theory: Objectivist and Constructivist Methods." In *Handbook of Qualitative Research*, edited by Norman K. Denzin and Yvonna S. Lincoln. Thousand Oaks, CA: Sage, 2000.

Collin, Matthew. *Altered State: the Story of Ecstacy Culture and Acid House*. London: Serpent Tail Publishers, 1999.

Crosby, David and Carl Gottlieb. *Long Time Gone*. New York: Doubleday, 1988.

Cummings, Sue. "Welcome to the machine:' the techno music revolution comes to your town." *Rolling Stone*. April 7 (1994):15-16.

Denzin Norman K. *Interpretive Ethnography*. Thousand Oaks, CA: Sage, 1997.

Frith, Simon. *Sound Effects*. New York: Pantheon, 1983.

Hitzler, Ronald. "Pill Kick: the Pursuit of "Ecstacy" at Techno-Events." *Journal of Drug Issues*. 32, 2(Spring 2002):459-65.

George, Nelson. *Hip Hop America*. New York: Penguin, 1998.

Jameson, Frederic. *Postmodernism: Or, the Cultural Logic of Late Capitalism*. Durham, NC: Duke University Press, 1991.

Jones, LeRoi. *Blues People*. New York: Morrow, 1963.

Kerouak, Jack. *On The Road*. New York: Viking Press, 1957.

Kotarba, Joseph A. "Rock 'n' roll music as a timepiece." *Symbolic Interaction* 25, 3 (2002a):397-404.

———. "Baby Boomer Rock 'n' Roll Fans and the Becoming of Self," in *Postmodern Existential Sociology*, edited by Kotarba, Joseph A. and Johnson, J. M. Walnut Hills, CA: Alta Mira, 2002b.

———. "The Rave Scene in Houston, Texas: An Ethnographic Analysis." Report for the Texas Commission on Alcohol and Drug Abuse (October), 1993.

Levine, David. "Good business, bad messages." *American Health*. May 10 (1991):16.

Lull, James, *Popular Music and Communication*. Newbury Park, CA: Sage, 1992.

Market, John. "Sing a song of drug abuse: four decades of drug lyrics in popular music— from the sixties through the nineties." *Sociological Inquiry* 71, 2 (2001):194-220.

Pareles, Jon. "Heavy metal, weighty words." *The New York Times Magazine*. July 10 (1988):26-27.

Chapter Eleven

The Neverending Conversation: A Case Study of Rave-Related Internet Conversation and Drug Use

Ann Lessem

At the outset of this study three questions came to mind: Are raves and club drugs inextricably intertwined? Do ravers use the Internet as a means to exchange information about these drugs? Does this communication increase drug usage?

Basic background research confirmed that an increase in the use of club drugs[1] had coincided with a leap in the popularity of techno music and raves (Hitzler and Pfadenhauer, 2002). Today, the common assumption is that raves and club drugs go hand in hand. For example, the following three definitions of raves from online slang dictionaries (representing three different continents) characterize them as places where youth gather to dance and take drugs.

> [Raves are] a large dance music party. The term originated in the 1950s and since the late 1980s in Britain has become synonymous with the "house" and "techno" dance scene. These raves are often illegal gatherings with much drug taking, especially "ecstasy" (British Dictionary of Slang, 2003).

> [Raves are] a dance party held in a large dance space, typically a warehouse or an outdoor clearing, and at which amphetamines are generally taken by the attendees. (Macquarie Dictionary Book of Slang, 2003 [Australia]).

[Raves are] all night dance parties frequently designed to enhance a hallucinogenic experience through music and lights (National Drug Control Policy, 2003
[United States]).

Just as more and more youth have attended raves and used club drugs since the
1980s, so too have more and more youth experienced the Internet and computer
mediated communication. A 2003 Mercury News and Kaiser Family Foundation
survey of more than eight hundred youth between the ages of ten and seventeen
found that over half used instant messaging or chat rooms at least once a week
and that one in four relied on "chat," instant messaging, or email as the primary
way they kept in touch with their friends (Plotnikoff, 2003). Because youth often
attend raves, often use drugs, and often use the Internet as a communication
channel, there is a widespread belief that a dangerous connection exists between
these venues, i.e., that youth who attend raves will use drugs and will communicate drug information over the Internet to other youth who attend raves, thereby
increasing drug use amongst youth who will attend still more raves and who will
communicate more drug information over the Internet, thereby increasing drug
use . . . ad infinitum.

However, very little research has been conducted regarding the extent to
which this cycle of raves, drug use and Internet communication about drugs actually occurs. To some extent this void can be explained by the difficulty in determining whether participants in any computer-mediated conversation about
drugs are actually drug-using ravers—even if they are conversing on a website
that, by its name alone, would lead one to believe that those very individuals are
the target audience. On the other hand, the problem of identifying online ravers
becomes much less onerous if the website being studied is devoted specifically
to raves. These sites typically serve as (a) a place to advertise future raves, (b)
an open space for composers of rave music to express their creativity, (c) a virtual photo album of past raves, and (d) a place where ravers can discuss almost
any topic of interest (Gibson, 1999). The question is, then, do drug-related conversations occur on these rave sites and if so, how do they impact drug use?

A cursory Internet search via a popular—although unnamed—search engine
for rave-related websites yielded over one thousand hits. Therefore, the first step
in this investigation was to devise a means for screening these sites to find one
that might be suitable for more intense scrutiny. To this end, the following process was used. First, rave-related websites that did not provide a forum for discussion were eliminated from consideration. Second, the remaining websites
that strictly regulated and controlled discussion topics were also eliminated. The
remaining sites were then scanned to discover if participants engaged in regular
conversation about a broad range of topics, i.e., conversation that might include
references to drug use. And finally, the few remaining sites were then monitored
for several weeks to determine if general discussions actually did include some
references to drugs. One rave-related website, which will be called *The Rave
Site*, emerged as having all of the required characteristics and the added benefit
of a totally unstructured conversation that site users challenged each other to

maintain for as long as they possibly could—"the Neverending Conversation."

The Rave Site was formed in 1999 and had over one thousand two hundred active members when it was observed for this study. The stated purpose of the site was "for people to come together in a positive environment to share their experience and knowledge related to dance music."[2] Rules of the site were very straightforward and included the following:

1. Play nice, treat each other with respect;
2. If you can't play nice, stay out of the sandbox;
3. Do not reply to abusive messages;
4. Keep on topic—post new topics in the correct area;
5. No useless posts;
6. No posting of personal information;
7. No warez,[3] copyrighted files, passwords, etc.;
8. No pornographic images;
9. Multiple aliases are frowned upon and may lead to removal; and
10. We reserve the right to update and change these rules at any time, for any reason.

The choices of discussion forums included the following:

- Dance Music Forum—for discussions specifically related to music, DJs, artists, record labels, clubs, clothing, publications, parties, raves, etc.
- Community Chatter—for general discussions with other members about any topic of interest, not necessarily related to music or dance.
- DJ Forum—for discussions with specific DJs about favorite styles of music, mixing techniques, places to spin, etc.
- Artist Forum—for technical discussions with other dance music artists.
- Marketplace—for buying, selling, and trading equipment and production hardware.
- Event Listing Board—for posting upcoming events.
- Politics and Current Events—for discussions about what's going on in the world.
- Health and Spirituality—for discussions about issues related to keeping fit, staying healthy, and living a long happy life.
- Dewdrops Garden—a place for women to discuss their point of view.

"The Neverending Conversation" began as a thread in the "Community Chatter" forum on January 31, 2001, and is still active as this article is being written. The following was the first post.

> I noticed how the site makes a new page for a topic after like 20 posts or so . . . What if there were like 1000s of posts for one topic? Hehe! I think we should start a new forum that is just entitled "The Neverending Conversation!"

For purposes of this research the conversation was monitored from its inception through March 7, 2003. During this time there were over 5200 posts and more than 170 individual participants. Approximately 90% of these individuals were male. The self-reported ages of the participants ranged from 15 to 103.

Patterns of Talk

Because the conversation continued for such a long period of time it was possible to quantify several factors. First, the posts were analyzed to determine whether participants were more prolific on certain days of the week. Second, the posts were analyzed to determine the time of day (or night) during which the participants were most prolific. And third, the volume of conversation over the entire research period was analyzed to determine the peaks and valleys of the discussion.

A seemingly reasonable assumption was that more conversation would take place over the weekends when participants would theoretically have more free time. However, this was not the case. Nearly one quarter of the conversations (24%) took place on Wednesdays, but only 17% of the conversation took place during the weekend—11% on Saturday and 6% on Sunday. 17% of the conversation also took place on Mondays and on Thursdays, thus further negating the assumption that weekends would be the most prolific. One explanation for the lack of weekend communication was that individuals who participated in the conversation could actually have been attending raves over the weekend.

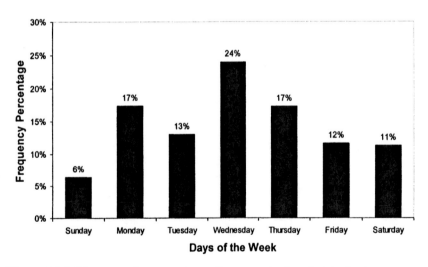

Figure 11.1. Illustrates the percentage of communication that took place on various days of the week.

Given the general inclination of youth to sleep late and stay awake late, another seemingly reasonable assumption was that most conversations would take place in the late evening or very early morning hours (wee hours). In order to determine the actual times conversations took place it was first necessary to pinpoint the location of the server on which the conversations were time-stamped. Numerous references in the conversations indicated that many of the ravers lived along the east coast of the United States. The posts of these ravers were examined for references to the time the raver was posting. Fortunately there were several instances in which this occurred. The raver-indicated post time was then compared to the time-stamp being placed on the post by the host server. This methodology confirmed that the host server was located in the Eastern time zone. Because so many of the participants were also from the Eastern time zone, it was assumed that the automatic time-stamp was a reasonably accurate indication of the actual post times. For purposes of analysis, the "day" was divided into four six-hour time blocks. These were:

- Morning (6:00a.m.—12:00p.m.)
- Mid-day (12:00p.m.—6:00p.m.)
- Evening (6:00p.m.—12:00a.m.)
- Wee hours (12:00a.m.—6:00a.m.)

The actual times of posts conformed somewhat to the original assumptions. The conversations were equally divided between the wee hours time period (27%) and the mid-day time period (27%), and conversations during the evening period were only one percentage point lower (26%). As expected, the fewest number of posts occurred in the morning (20%). Figure 2 illustrates the time of day individuals participated in The Neverending Conversation.

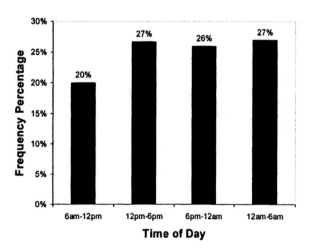

Figure 11.2. The time of day during which individuals participated in The Neverending Conversation.

The duration of the study made possible a third analysis of the conversational patterns—volume over time. The Neverending Conversation was monitored for 109 weeks. However, the amount of talk (calculated as the number of posts) was not consistent during this time period. The conversation ebbed and flowed between a high of over nine hundred posts in one week to weeks in which there were no posts. The discussion volume, by week is depicted in Figure 3.

Volume by Week

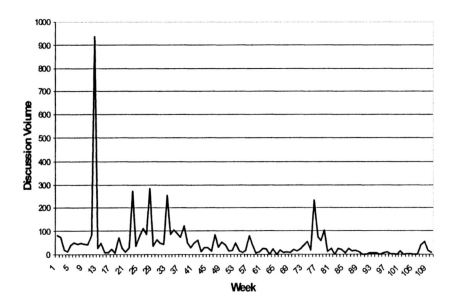

Figure 11.3. Volume of discussion, calculated as the number of posts by week.

In addition to the highest spike of posts during week 12, there were a series of moderate spikes (over two hundred posts) during weeks 24, 28, 34, and 75. There did not seem to be any obvious external reason for these periodic escalations in activity such as holidays, extended school vacations, or cycles of the moon. But an examination of conversation content quickly solved the mystery. In all instances, these spikes in posting were due to one of two closely related factors. The first factor being an attempt by participants to reach some milestone in the number of posts made to the conversation. The second being an attempt by one of the conversation participants to reach a personal milestone related to the number of posts they had made to the site. For example, the week of nine hundred posts was an attempt on the part of one raver to have more than two thousand posts on *The Rave Site*. Other participants in The Neverending Conversation eagerly helped him to reach this mark as they engaged in various forms of nonsense postings, which they called "cheating." These included tactics such as

- A series of short phrases and incomplete thoughts posted by the same individual many times consecutively in order to complete one thought (for example—Post #1—I'm going; Post #2—to turn; Post #3—up the; Post #4—volume);
- A series of incomplete phrases or thoughts that would be "batted" back and forth between two or three participants (for example, Post #1, Raver #1—Willy went to the store; Post #2, Raver #2—to buy some pork and beans, but when he got there; Post #3, Raver #3—the shelves were bare . . .);
- A series of consecutive posts used simply to create a long list of similar items, such as a list of vegetables in which the first post says "carrots," the next says "peas," the next says "broccoli," etc.:
- A series of posts in which the same word or words were repeated over and over and over; and
- A series of posts of nonsense combinations of letters and/or numbers.

In other words, the high volume of posts during some weeks was unrelated to conversational topics and was simply an artifact created as the conversation participants played with increasing the number of posts. On the other hand, slight-to-moderate increases in the number of posts typically occurred shortly after a new topic was introduced and tended to fall off after the topic had been discussed for several days. Over the course of the study the participants in The Neverending Conversation challenged themselves again and again to come up with new topics that would be of interest. Tactics they used to do this included:

- Starting fantasy stories that would end in mid-sentence with the expectation that another participant would add more;
- Starting discussions about musical preferences and favorite DJs;
- Starting discussions about sex and sexual fantasies; and
- Starting discussions about things they did because they were bored (such as watching television, surfing the net, watching people, making up stories about each other, etc.).

Content of Talk

The analysis of easily quantifiable items such as days of the week, times of the day, and volume of discourse were interesting, however they were not instrumental in answering the original research question—"do drug-related conversations occur on rave-related websites and, if so, how do they impact drug use?" Quantitative analysis was abandoned in favor of an in-depth ethnographic analysis of conversational content. The premise underlying this analysis was that the parts of the conversation could only be understood in the context of the entire conversation (Lepper, 2000). Therefore, the analysis focused on all dialogue rather than only the drug-related exchanges.

Communication Anomalies

The 109-week duration in which The Neverending Conversation was monitored involved an on-going analysis of over 5200 posts. While this was an overwhelming amount of conversation in its own right, the complexity of the analysis was compounded by the unique nature of discourse that went on between participants. As mentioned above, some of the conversation was undertaken as a means to surpass arbitrary numeric milestones in number of posts rather than as a means to convey information or to interact with other participants. In addition, other conversational characteristics were made possible because of the media (the Internet) in which the conversation/communication took place. These anomalies had several, often dichotomous characteristics.

Synchronous/asynchronous
As with most web-based chat, the conversation often simulated face-to-face communication in that multiple individuals were engaged in synchronous exchanges. A short example of this synchronous "talk" was apparent during the following exchanges in which participants were trying to reach a milestone number of posts before 3:00a.m.

> [Raver 1] That would be like 1 minute per page . . . 1 post every couple of
>   seconds
> [Raver 2] Yes
> [Raver 1] Poop
> [Raver 3] 2.59:13
> [Raver 1] I just called the "shit" "poop"
> [Raver 2] Um, forgot where I left off
> [Raver 4] We need more power capn, I can't keep her together!
> [Raver 1] We didn't do it.
> [Raver 3] We missed it, it's 3:00:24
> [Raver 2] So close.
> [Raver 1] 3:05 . . . we can do it before then

However, these exchanges were written rather than spoken and the fact that they were written, and that this written content was maintained by the website in an easily accessible manner, made it possible for individuals to retrieve what was written and read it. The following posts, in which the posters had been absent for a period of time or had just joined the conversation, were examples of exchanges that implicitly acknowledge the asynchronous nature of the conversation either through comments about what had previously transpired or comments in which new posters were directed to search the archives so they could catch up.

> Well well—I have been off for a few and I see that the story has not progressed much.

> I stop posting for a whole 3 days and you guys stop the story, what's up with that?

> I'd explain it to you . . . but uhhh, I think I agree with [Raver name] . . . it'd be funnier to have you go back and read the last 84 pages!!!

> For the newbies I would suggest going to pages 50-75

Inclusive/Exclusive

The Rave Site forums could be accessed by anyone who had a computer and who knew the web address. Certain aspects, such as archives of conversations, were available to everyone regardless of whether they were a "member" of the site. Other aspects, such as being able to actually participate in the conversations, were restricted to members. However there was no screening process or fee for this membership. Therefore, anyone who so desired could take full advantage of website participation. This spirit of open communication between all individuals interested in raves was one of the hallmarks that some ravers were proud of. They often characterized themselves as practicing *PLUR*—peace, love, unity, and respect—with their fellow ravers. The following example is just one of several instances where existing participants in The Neverending Conversation explicitly welcomed a new individual into their conversation.

> [Raver 1] I used to be afraid to post stuff on here
> [Raver 2] The first couple of months we were all that way. Takes awhile to get
>   used to things.

However, even though The Neverending Conversation was available for any and all comers, certain aspects belied this inclusiveness. Some of the individuals engaged in the conversation acknowledged that not everyone might feel like they really belonged.

> This should be changed to the "BBS resident"[4] chat conversation. Nobody under 50 posts has posted on this conversation. It has been all BBS residents. I wonder if that says anything about us?

And some individuals who joined the conversation immediately knew that they were not going to participate any further.

> This is the first time I have ever been in here . . . I won't be back. My life is pretty damn boring but good god!

And unknown others, who may have lurked[5] but who never joined in, may have been put off by the "in-house" references and slang that permeated the conversations.

> Shit i got off the bazzanklesworth that night! I was hip hopin my aristocratic styles with some rappin and I was scratchin in some little mermaid theme song over the Who's the Boss intro music! Man they were screamin my name and throwin glow in the dark rubber sheets for bed wetting at me!

Profound/profane
The participants in The Neverending Conversation often reflected on their life and their place in the world. In some instances these reflections were quite profound.

> A sea of infinite possibilities, all with their own neverending story, which path do we take?

> Never underestimate the power of stupid people in large groups

> Always look out for the guy behind the guy in front of you

On the other hand, some of their reflections and comments were very sexual, and might be considered mildly profane.

> ~whew~my ass sure hurts . . . damn, must be cause I just got raped like a 10 year old boy . . . and that really sucks considering I'm 23

> Although I may be the only girl posting I may not be the only one giving head

> If you sit on your own hand long enough you won't even notice that it's not some very attractive women using you

Fantastic/mundane/surprising
One of the other unusual things about the posts in The Neverending Conversation was the frequency of fantasy conversation. The participants relished coming up with fantastic stories, contributing to each others fantastic stories, and/or just encouraging each other to make their stories even more fantastic.

> and the pickles were in an uproar . . . Who could save the ceiling fans who smelled of taco bell? Only a virgin platypus with the persistence of a candy kid begging for a dollar could turn the peanut butter back into mayonnaise.

> But then a giant one armed tree came out and started to throw doughnuts at the cop and the cop chased after the flying pastries and they all managed to delete the virus and korg began to play the phatest beats you have ever heard and they all continued on their journey down the escalator but then . . .

> ATTENTION as of [date] there will be a new story in the works!!! This story will contain the long awaited sequel to the original Raver Wars. Also look for

all new characters and events to take place including the destruction of The Dope Star and the introduction to DJ Darth Vapor Rub's son named "lil Luke Inhaler." Ravers Wars 2 "The Empire Smokes."

In sharp contrast with the fantastic, many of the posts were about almost nothing at all as the participants struggled to keep the conversation going and to fulfill the initial challenge of having the longest single conversational thread on *The Rave Site*.

> We're up to 12 pages already . . . WooHoo!!! Only 88 more to go before we reach a hundred pages
>
> My fridge smells like pizza

And there were other times when the conversation encompassed topics that one might not expect to hear/read on a rave-related website.

> [Raver 1] I would like to invest in a low risk variable annuity and live life like I always do, and take a monthly payment from the interest
> [Raver 2] low risk = lame. You don't make money that way man. It sounds weird coming from a 15 year old but I'm a loser who's been reading about money a lot the past year. I think if I had a mill . . . I'd invest it on setting up a studio then I'd invest the rest in real estate

Communication without words/words without communication
The web-based medium enabled the conversation to be visual as well as verbal. In some instances the participants simply posted pictures and/or animated cartoons to convey their thoughts.

Figure 11.4. An example of an animated cartoon used to convey thoughts.

On the other hand, the participants often interacted with The Neverending Conversation as though it was really not a communication medium at all. Sometimes they treated it like it was a game. This game-like attitude was apparent in their quest for post-related milestones.

I wanna have a whole page to my own posts

Or sometimes they would find themselves the only one online at a particular time and would use the forum to talk to themselves.

Well I guess I will continue this conversation by my damn lonesome self as most anybody that uses this board is in florida and therefore is at DJ WARS

Ok well since I am all alone this morning I figured I would start a conversation with my self and see what I get out of it!

Drug-Related Discourse

But did they talk about drugs? The simple answer was "yes." Mingled within the multiple types and styles of communication were numerous references to drugs and drug use. However, the fact that participants made reference to drugs was not enough to completely answer the research question—"do drug-related conversations occur on these rave sites and, if so, how do they impact drug use?" It was therefore necessary to further analyze the ways these ravers talked about drugs.

By far the most common form of drug-related communication was the use of drug slang. This slang typically served as a basis for group cohesion, i.e., only a member of their group, or of a similar group, would understand the slang terminology. Examples of the drug slang used by participants in The Neverending Conversation included (but were not limited to):[6]

- Ballin—using cocaine
- Beans—MDMA (methylenedioxymethamphetamine);
- Buddha—marijuana laced with opium;
- Crippy—high quality marijuana laced with crack;
- Hydro—marijuana grown through hydroponics;
- Lemonade—heroine;
- Macaroni and cheese—a $5 pack of marijuana and a dime bag of cocaine;
- Pork and Beans—a food item containing marijuana;
- Salad—marijuana;
- Special K—ketamine;
- Zen—LSD.

And even though drug references permeated many conversations, they were typically attempts at humor and sarcasm.

[Raver 1] lol[7] . . . typical crackhead response too
[Raver 2] Yeah, but I'm not a typical crackhead . . . lol
[Raver 1] lol . . . yeah but that's cuz you cut yours with Ajax!
[Raver 2] Hey I don't cut my crack with Ajax . . . I use Comet bathroom
        cleaner . . . get it right . . . lol
[Raver 3] hahahahahaha you guys are nutz!
[Raver 2] I'm a person. Nuts grow on trees . . . lol
[Raver 4] Doesn't everyone use Comet these days?
[Raver 2] Yeah that's the magic ingredient in crack. It gives it a little edge . . .
        lol
[Raver 1] naw man . . . it's all about the Ajax! Hehehehehe

I'm no genius . . . but I'd say that looks a lot like the molecular structure of
Methylendendioxymethampethamine[8] . . . hehe either that . . . or a wiener. I always seem to get those 2 confused

Drug-related references (again typically humorous or sarcastic) were often the
basis for their fantastic stories.

Gumbee says, "But he took my bowl of pork n' beans . . . not fair!" Ariees
says, "Sharaz, are you the one taking all the pork n' beans?" Sharaz says, "yes
cuz I heard BMike put acid on all of it." So everyone started tripping out and
the DJ ended up playing contemporary jazz all night at the rave

So Princess "Come On I Wanna" Leah lived there because it was rent free and
she could help out selling lemonade with GHB in it at her stand on the side of
the road

Sometimes the participants did indeed speak very seriously about drugs and
drug use. However, there was not one reference in the five thousand two hundred posts analyzed during this study in which serious drug-related conversations encouraged drug use or provided information about where/how drugs
could be purchased and/or who was selling them. Surprisingly, when the drug-
related conversations turned serious the ravers from The Neverending Conversation were either opposed to, or ambivalent about, drugs. This, despite the fact
they accepted that drugs and raves seemed to go together and often acknowledged their own drug use.

Drugs are a part of the scene, whether you like it or not. They have and will be
used. I don't use them anymore, and I think it's unnecessary, but who am I to
say that a person shouldn't take a drug—all that matters to me is that they are
enjoying the music

What I found funny was that no matter what tent[9] I was in I smelled hydro, not
reg weed, but hydro. That's [city name] for ya. I kind of felt that I would have
had a better time if I was fucked up, but I don't regret not doing it

What's up with all these kids and all these drugs? Doing it to the point where you don't even know what party you're at or who's music you're listening to! Now that's messed up! I know this has been going on forever and has been talked about sooo many times but I just had to say something. ITS WAKKKKKK! DO IT AT HOME

I'm not anti-drug, I'm anti-stupidity. If DJs are going to complain about the drug problem . . . don't play to the people on X

There were wayyyy too many rolling candy raver retard kids. I didn't realize those things existed anymore

It's one thing to smoke the Buddha all day and actually enhance the depth at which you hear good music, but to eat pills and get stupid and then just love everything just because you ate some shit . . . That's not enhancing anything, that's brainwashing yourself into being stupid

Conclusions

The Neverending Conversation is just one of potentially thousands of web-based, rave-related forums. Therefore it is impossible to say with total certainty whether the attitudes expressed by the ravers participating in The Neverending Conversation were representative of the norm. Regardless, this conversation has brought two very enlightening items to the forefront.

The first revelation is that there are multiple types of ravers. As was pointed out in the more serious discussions, some ravers attend raves for the music and are not interested in drugs. Therefore, in order to more thoroughly understand raves and their relationship to drug use we must more thoroughly understand the heterogeneity of the ravers. To characterize all ravers as the same would be tantamount to characterizing all people who attend a county fair as being the same simply because they chose to attend the same event.

The second revelation is that rave-related, Internet-based conversations about drugs are not necessarily endorsements of drug use. Even though Ravers exchange information and communicate about drugs as they converse over the Internet, the majority of these interchanges are playful rather than serious discourses about how, when, and where to secure or use drugs.

In The Neverending Conversation the ravers referred to, and talked about drugs in four different ways: (1) they used drug related slang as a conversational means to identify themselves and others as members of the same group; (2) they used humorous references to drugs as a means of entertainment, much as comedians often make reference to things during their acts because they are funny, not because they actually do these things when they are off stage; (3) they used drugs as a prop and/or an event in fantasy stories, much as other props and events are used by authors of fiction; and (4) they talked seriously about drugs and drug use.

While raves and club drugs might overlap in several ways and rave-talk might include many references to drugs, the connection between raves and drug use does not seem to be as inextricably intertwined as is generally assumed. This is in sharp contrast to many stories about raves that convey alarm, not only about drug use, but about raves being associated with "dens of doom," political power plays, subversiveness, transgression, etc. Christopher Stanley (1995) characterized raves as one of three dissenting narratives of youth that occupied deregulated space and were therefore deemed "wild zones"—joy riding and computer hacking being the other two. However, what has been learned through The Neverending Conversation raises doubts about the inherent dangers of raves and rave-related Internet conversations. Contrary to the assumption that ravers' communication via the Internet is a peril that must be curtailed in order to protect youth, is the insight that Internet sites such as *The Rave Site* and communication forums such as The Neverending Conversation are portals through which the lives of youth who attend raves can be more thoroughly understood.

Notes

1. Club drugs typically include Ecstasy (MDMA), Lysergic Acid Diethylamide (LSD/Acid), Rohypnol, Methamphetamine (speed), Ketamine, and Gammahydroybutyrate (GHB). These particular drugs distinguish themselves from their predecessors (e.g., heroin and cocaine) in their most common methods of administration (oral consumption, snorting, or smoking are more common than injecting) and their effects (reducing social inhibitions, increasing sensory perceptions, and causing a feeling of being at peace with the world).

2. As per the FAQ section of the website

3. "warez" refers to software that facilitates illegal access to protected computer sites and files

4. "resident" is a slang designation given to members of the site who have posted an assigned number of times

5. "lurkers" are those who observe but who do not actively participate

6. Definitions are per several online drug slang dictionaries such as Street Terms: Drugs and the Drug Trade located at http://www.whitehousedrugpolicy.gov/streetterms/ByAlpha.asp?strTerm=A and Erowid Drug Slang Vault located at http://www.erowid.org/psychoactives/slang/slang10.shtml#S

7. "lol" is a type of Internet short hand, and means "laughed out loud"

8. This comment was made in reference to a molecular-looking picture that had been posted by another participant in the conversation.

9. Referring to multiple tents that had been set up a the Rave this participant had just attended

References

British Dictionary of Slang (online). <http://www.peevish.co.uk/slang/search.htm> (2003).

Drug Facts, Office of National Drug Control Policy (online)<http://www.whitehouse
    drugpolicy.gov/streetterms/default.asp> (2003).
Hitzler, R. and Pfadenhauer, M. "Existential strategies: the making of community and
    politics in the techno/rave scene." In *Postmodern Existential Sociology* Ed. Kotarba
    and Johnson. Altamira Press: Walnut Creek, 2002.
Gibson, C. "Subversive sites: rave culture, spatial politics and the Internet in Sydney
    Australia." *Royal Geographic Society* 31.1 (1999): 19-33.
Lepper, G. *Categories in Text and Talk.* Sage. London, 2000.
Macquarie Dictionary Book of Slang (online) <http://www.macquariedictionary.com.au/
    p/dictionary/slang.html> (2003).
Plotnikoff, D. "Wired kids." *The Mercury News,* Aug. 5 2003. MercuryNews.com.
    (2003).
Stanley, C. "Teenage kicks: urban narratives of dissent not deviance." *Crime, Law and
    Social Change* 23 (1995): 91-119

Chapter Twelve

Using Popular Music to Interpret the Drug Experience

Shawn Halbert and Joseph A. Kotarba

A common truism in our popular culture today is that rock 'n' roll music is the soundtrack for many people's lives. Rock 'n' roll music specifically, and the varieties of popular music derived from rock 'n' roll more generally, have served this function ever since their inception in post-World War II America (Kotarba 2002). Three generations of Americans—and Europeans, for that matter—have used youth-oriented popular music to make sense of relationships, parents, authority, leisure-time activities, work, fun, and life itself. In effect these understandings have been critical to the construction of the myriad identities people have employed throughout their everyday lives.

As an art form, rock 'n' roll music thrives and owes much of its popularity to the way it is able to help people make sense of otherwise ineffable experiences. Rock 'n' roll helps young people make sense of God, for example, by means of the genre of Christian heavy metal (Kotarba 1994). Fans ranging from teenagers to adults have used rock 'n' roll to make sense of feelings ranging from ecstasy to painful loss. This cognitive process is especially important in the formation of identities that are defined through visceral or transcendental experiences. The ineffable experience in question here is drug use relevant to a wide range of popular music styles, especially techno or rave music.

Regarding drug use specifically, the more traditional sociological approach to music and meaning is to analyze the social, cultural, legal, aesthetic, and temporal contexts within which recreational drugs are used and which produce drug

197

experiences. These reports often focus on the "top-down" processes of labelling and categorizing along with evaluations of social factors that may or may not determine a preference for drug use. This focus often results in a common neglect of the everyday work that individuals undergo to construct their identities as drug users. A minority of studies have employed this latter approach, concentrating instead on the "bottom-up" processes of identity formation to demonstrate the emergence of the self-as-drug user. Howard Becker (1953) wrote one of the earliest statements on this process. He explained how face-to-face interaction with others is the primary source of meaning for the initial experience of marijuana use. By themselves, the effects of marijuana are neither pleasurable, painful, nor regrettable. Others, especially those also engaging in marijuana use, provide interpretations of the experience that can be definitive for the neophyte.

Becker was especially poignant in explaining that all aspects of the self, even understandings of visceral experiences, are mediated in social interaction. This is important in that it places emphasis on individuals' reference groups as influences on the formation of their identities. Becker later demonstrated how this becomes problematic when the availability of reference groups is limited as a result of social stigma (1963). The presence of sanctions against drug use necessitates a level of anonymity that presents a conundrum for individuals who are struggling to understand their selves-as-drug users. For many, the bifurcated discourse of deviant/non-deviant compels them to suppress this aspect of their identity as they work diligently to manage their identities against the consequences of potential stigma (Goffman 1963). This suppression can result in maladies such as anxiety and depression, and has promulgated the discourses of powerlessness that are prevalent in rehabilitation programs such as Alcoholics Anonymous (Denzin 1987). Furthermore, the need for anonymity significantly reduces variation in available reference groups. As a result individuals wishing to make sense of their selves-as-drug users must associate with groups that have some visibility but are, more often than not, categorized by their deviant behavior. Thus individuals come to associate their selves-as-drug user with other deviant behaviors and identities.

The virtual world of the Internet could presumably alleviate some of these barriers to a positive construction of the self-as-drug user. The Internet offers many the opportunity to play with multiple selves in a safe, anonymous environment (Turkle 1995). The delimiting of spatial and temporal restrictions and the anonymity provided by the Internet fosters the development of myriad communities whereby individuals can draw resources of intersubjective understandings necessary for self-construction (Surrat 1998, Adams 2003). Zhao (2005) demonstrates the effectiveness of the Internet in developing the same resources for identity formation that we find in others. He explains that interactions can occur in "tele-copresence" with others, allowing individuals to derive understandings of their selves through the eyes of mediated intimate strangers or a generalized other of limitless proportions (Altheide 2000).

While the ability to remain anonymous and locate references groups with considerable variability reduces many of the problems users face in constructing their selves-as-drug users, the Internet presents potential challenges of its own. As Becker (1953) understood, the becoming of the self-as-drug user requires attributing meaning to the otherwise ineffable experiences of such practices. These are not easily conveyed through a strictly textual medium. As a result individuals must construct and present their selves through stories (Holstein and Gubrium 2000) that require an extensive use of mutually understood references to cultural artifacts, such as popular music, as metaphorical resources (Lakoff and Johnson 1980).

This understanding of the Internet helps us to apply Becker's ideas to current drug issues. Since Becker's seminal analysis, popular music, drugs, the audience, and communications media have all evolved. The audience is composed of those people who listen to techno, rave, and other currently popular styles of popular music by means of pre-recorded CDs, downloads from the Internet, and by means of DJs in dance clubs. Of special interest is the way they talk about drugs and music. They share stories about drug experiences made sense of by means of discourse on the Internet. A very popular address to host this discourse is DrugSite. Becker described the impact of the face-to-face marijuana subculture. We will demonstrate that the same socially constructive process occurs today with designer/dance drugs, but that the subculture in question operates online. Further, we will examine the actual music used to add meaning to the drug experience.

The research question guiding this study is: what are the various ways participants in DrugSite use musical phenomena to make sense of their drug experiences? The method involves a content analysis of messages posted in rooms and threads over a one year period (2004-2005). Analysis followed the grounded theory style: comparing data as it is collected, developing theoretical categories as they emerge from comparison, and using these categories to guide subsequent collection and refine emerging theory (Glaser and Strauss 1967; Charmaz 2000). The excerpts from actual posts are literal.

Music and Drugs

Musical topics appear consistently throughout DrugSite's posts. They generally fall within six categories: musical style used metaphorically to describe drug experiences; specific songs used metaphorically to describe drug experiences; lyrics used to describe drug experiences; performers associated with drug experiences; DJ styles associated with drug experiences; and quitting drug use.

Musical Style

Certain musical styles "feel like" or remind people of the times that they did certain drugs. The styles in questions are conventional popular identifications.

The first example demonstrates how an author draws a comparison between the feeling of doing speed and style of heavy metal music known as thrash:

> We got some crystal from some guys at the show. My first time and I have to say . . . whoa! The whole place turned into one big thrash concert heart beating like the double bass and the noise of the crows blurred and accelerated like the thrashing of a highly distorted axe.

The second example is an author who refers to a first time doing pills—apparently barbiturates or valiums. She was previously "rolling" on ecstasy and she drew on past experiences with a certain form of music to describe the feeling:

> I'll never mix the two again. It reminded me of the way I always felt listening to goth music: dark, angry, etc . . . I didn't like the feeling I got when I heard that stuff and I certainly didn't like how I felt last night!

A third example is an author whose post refers to a genre of music, trance, often associated with ecstasy. The author had experience with trance before doing "E" and was told that the music reflected the experience. He never understood that until he tried ecstasy. His experience matched the music in his mind:

> I have to say I loved House and Trance before ever doing anythin. My friends kept telling me I wouldn't understand until I rolled. I broke down and did it last weekend and now I know!!!!. Trance is written for rolling, I felt like I had done it before!!! But I hadn't!!! The music matched the feeling!!!

Songs

Many authors post messages that refer to specific songs. They typically reference songs by artist and title, with a brief introduction explaining why they refer to that particular song. Very rarely did the author go into detail regarding the specific qualities of the song. Many of the postings invited the reader to listen to the song, as if the author could not verbalize the feeling and was therefore leaving it up to the reader to experience the song to get their point.

For example, one author describes an experience with mushrooms in detail. He thoroughly explains communicable experiences such as the visuals, the sounds, the activities and the experience, yet when he tries to describe the physical feeling he can only draw analogies. Instead of going into too much detail, he simply refers to a song—*The Warmth* by Incubus—and tells readers that they can get what he's saying from that. The entire narrative follows, and the section where he points out the song is in bold:

> Let my first say I have done mushrooms probably 50+ times and this was one of the craziest trips I have ever had. Also the weirdest thing about it was I only

ingested about a half 8th or a little more.

I'm a senior in high school and my finals just ended and I did really well on all them, my dad just had a really successful surgery and is already feeling a hell of a lot better. I am in a good mental setting and I had wanted to take mushrooms again for awhile after 5 month hiatus from them.

So me and my buddy P were both gona be tripping but our friends J, E, and D, were just going to chill around and smoke the bud we had all pitched on early which was about a quad or so.

Everyone starts smoking but I choose not to because I wanted to go into the mushrooms world sober. I ingest my mushrooms along with P and then down a glass of orange juice to get the taste of mouth.

P's dad calls and tells him to go to feed the his horses so I am left with J, E, and D we began talking and just chilling and they are smoking bud out of the newly broken bong(it sill works though!) I soon began to feel the first effects of the mushrooms kicking in I decide to go for a walk and look for P. I find him feeding his horses and ask him if he feels it yet and says he defiantly does. We take a walk around his property and talk about different things and smoke some cigarettes. Everything is so sharp and clear it was very a nice come up.

P has a couple of more things to do around his house so I go back to his room and watch J, E play halo 2 for awhile. This is where it starts getting kind of hazy but I will try to explain everything as best I can. Time kept slowing down throughout this whole trip. It had prob been an 40 min since I took the shrooms and it had seemed like forever and now I knew I was tripping hard cause I could see patterns forming everywhere around me and the colors were just amazing, I was seeing these beautiful greens, blues, reds, purples, and the whole time I was just smiling listening to random music.

At this point it's probably been about an hour and half into the trip and P is done with the things he had to do around his house and he was tripping balls. We all decide to go outside and smoke some cigs and look at the sky. Everyone now is heavily intoxicated on either mushrooms, or weed and loving it! I see the clouds moving in the sky and they are moving what looks to be a million miles and hour and forming all kinds of crazy ass animals and people. The craziest cloud I saw was a cat body with my head running through the sky.

We go back inside and I am tripping so hard all I could do was laugh. My buddy P unfortunately isn't doing to hot and is going to a very bad place. I felt very sorry for him but knew there was nothing I could do except tell him everything would be ok in a couple of hours and play some of his favorite music for him get him a glass of water. He went outside to be alone and we all let him. It's been about 2 hours now since ingestion(it felt like I had been tripping for hours though) and I am now peeking and my friends are smoking pot again so I decide to join them. Right when I exiled the hit I was transported to another level of tripping everything was moving so fast. I began talking to my friend J, but at the same time was having like 3 other conversations in my head with D and E and myself. This part of the trip was strange because I thought the weed would slow me down like it usually does but not this time.

At this point P finally comes back inside he smokes some pot with the rest of crew while I go lie down, I close my eyes and I am presented with the coolest CEV's I was seeing mushrooms everywhere around me and when I opened my eyes I was presented with more mushrooms but this time growing all over the room. I went outside and had a smoke and just admired the world I was in and felt truly blessed for living in such a beautiful town and having such good friends. I felt perfect in everyway at this point; but soon my sense of perfect ness was gone and all I felt was selfishisness because I was feeling so good but all around the world people were feeling so much pain. I thought about the people in Asia and how they must feel after loosing there friends and families and essentially their lives. I prayed for souls who perished in the tsunami and hoped they found it to the next life safely.

I went back inside and explained this to everyone and it felt as though all their energy was going it to me and they all felt the pain I was feeling. Again I went and layed down and was presented with more CEV's but this time I saw energy being created in my body. I t was a beautiful green color exploding over and over again. Those were the last of my CEV's and slowly the real world snuck back in. I was left with an azazing body high though that lasted or another hour or so.

This trip was one of the most amazing trips of my life and throughout the come down me and my friends laughed listened to music smoked bowls and just hung out and had a good time. I am really looking forward to taking mush-rooms again next weekend probably the same amount or a little less because I will be at Disneyland. Hope you enjoyed this report and sry for the length. I wish I could go into more detail but it still going through my head at a millions miles an hour. **If you wanna know what I was feeling, try listening to "The Warmth" by incubus it's a really cool tripper song and I listened to it a lot that night because it reflected what I was feeling.**

P.S. Also at some point everyone left to go pick some P's girlfriend up and I was left alone and saw this episode of Seinfeld where George's girlfriend has a doll that looks like his mother and it drives him crazy he even talks to it in a re-straunt. And for some reason Kramer and Georges dad (forgot his name) aren't wearing pants because they want to keep the crease in them and it was trick they learned from some symphony director. None of my friends believed me when I told them they said I was just tripping but I know it happened. Has any-one seen this episode?

Songs can also remind people of special and memorable highs, helping them to connect and interpret visceral feelings from the past with external experiences in the present:

The spacey feeling of "the division bell" by Pink Floyd takes me back to my smoking days

Finally, there is a sense among participants posting song titles that the readers know the songs and have shared the same experiences. While these postings share attempts to understand the feelings of drugs, they also infer an acknowledgement of a community of shared identity and understandings through what they take for granted. The following post demonstrates this in terms of the way the author infers that the reader should know the song and know how that song "feels:"

> What is dust like? "Amazing" by the Offspring . . . Nuff Said

Lyrics

There are many examples of sense-making activities that utilized lyrics throughout DrugSite. Themes within the lyrics can be divided into two categories: (1) the words describe a state of mind (whether they were intended to or not), and (2) the lyrics describe an experience that the author relates to from past drug experiences. The following are six examples of lyrics quotes and lyrics use. We place the participants' comments in italics and the lyrics to the song in normal font. We place singled-out lyrics in bold:

First example:
> Hey all!!! Tis been a helluva week for me. Looking for the weekend so we can drop some more and get lost. I'm posting for wishful thinking. They gotta be talking about dropping cuz that's what its like. Enjoy!!!

> **Tool—Lateralus**
> Black then white are all I see in my infancy.
> Red and yellow then came to be, reaching out to me,
> lets me see.
> As below, so above and beyond, I imagine,
> drawn beyond the lines of reason.
> Push the envelope, watch it bend.
> **Over thinking, over analyzing, separates the body from the mind.**
> Withering my intuition, missing opportunities and I must
> Feed my will to feel my moment drawing way outside the lines.

> Black then white are all I see in my infancy.
> Red and yellow then came to be, reaching out to me,
> Lets me see there is much more
> and beckons me to look through to these infinite possibilities.
> As below, so above and beyond, I imagine
> drawn outside the lines of reason.
> Push the envelope. Watch it bend.

> Over thinking, over analyzing separates the body from the mind.
> Withering my intuition leaving all these opportunities behind.

Feed my will to feel this moment urging me to cross the line.
Reaching out to embrace the random.
Reaching out to embrace whatever may come.

I embrace my desire to
feel the rhythm, to feel connected
enough to step aside and weep like a widow
to feel inspired, to fathom the power,
to witness the beauty, to bathe in the fountain,
to swing on the spiral
of our divinity and still be a human.

With my feet upon the ground I lose myself
between the sounds and open wide to suck it in,
I feel it move across my skin.
I'm reaching up and reaching out,
I'm reaching for the random or what ever will bewilder me.

And following our will and wind we may just to where no one's been.
We'll ride the spiral to the end and may just go where no one's been.

Spiral out. Keep going, going . . .

Second example:
    —these are my dropping balls get lost lyrics—

Peter Gabriel—Washing Of The Water
River, river, carry me on
Living river, carry me on
River, river, carry me on
To the place where I come from

So deep, so wide, will you take me on your back for a ride
If I should fall, would you swallow me deep inside
River, show me how to float, I feel like I'm sinking down
Thought that I could get along
But here in this water, my feet won't touch the ground
I need something to turn myself around

Going away, away toward the sea
River deep, can you lift up and carry me
Oh roll on through the heartland
'Til the sun has left the sky
River, river, carry me high
'Til the washing of the water, make it all alright
Let your waters reach me, like she reached me tonight

Letting go, it's so hard, the way it's hurting now
To get this love untied
So tough to stay with this thing, 'cos if I follow through

I face what I denied
I'll get those hooks out of me
And I'll take out the hooks that I sunk deep in her side
Kill that fear of emptiness, that loneliness I hide
River, oh river, river running deep
Bring me something that will let me get to sleep
In the washing of the water will you take it all away

Example three:

im in a hell good mood (it seems the songs i choose lately have all been in oposition to my moods) but this song always strikes me as amazing art. reminds me of my 'shroomin times.

Open—The Cure
i really don't know what I'm doing here
i really think i should've gone to bed tonight but . . .
just one drink
and there're some people to meet you
i think that you'll like them
i have to say we do
and i promise in less than an hour we will honestly go
now why don't i just get you another
while you just say hello . . .
yeah just say hello . . .

so I'm clutching it tight
another glass in my hand
and my mouth and the smiles
moving up as i stand up
too close and too wide
and the smiles are too bright
and i breathe in too deep
and my head's getting light
but the air is getting heavier and it's closer
and I'm starting to sway
and the hands around my shoulders don't have names
and they won't go away
so here i go
here i go again . . .

Example four:

beig vry drunk i apolgis now gfor typos and stuff . . . tis 12.07am and i **FINALLY FOUND** *the song i have heard for a while now htat i canyt sto singing . . . i heard thois in the car with twigz a fellow bluelighter todat and i said tis just lik my time on E. Like I was waiting for it to com. I finally found what it is called after heaRING IT IN SANITY . . .*

Artist: Evanescence
Song: Bring Me To Life

Album: Fallen

How can you see into my eyes like open doors?
Leading you down into my core
Where Ive become so numb

Without a soul
My spirit's sleeping somewhere cold
Until you find it there and lead it back home

-CHORUS-
[wake me up] Wake me up inside
[I cant wake up] Wake me up inside
[Save me] Call my name and save me from the dark
[Wake me up] Bid my blood to run
[I cant wake up] Before I come undone
[Save me] Save me from the nothing Ive become

I LOVE THIS SONG!!!!!!!!!!!! 😃😃😃😃😃ps.i ended up just putting thw
hole song in bold 😃

Example five:
this is quit lengthy and i apologize in advance.

last night was my first time to take shrooms. three other (experienced) friends
and i took 1/16 each as we were told these were strong. we chewed them and
swallowed with some iced tea at 11:15 pm.

around 11:30 i went outside to smoke, and i saw a little rainbow outlining the
moon. i wasn't sure if the shrooms were kicking in, and i thought i was making
myself see it because i was anticipating the effects. when i was done smoking,
the body load kicked in. we were in my room with dj sammy's "sunshine"
playing on repeat, and i lied down in my bed because i didn't want to move.
just then i started laughing at everything! i felt very giggly and could not stop
laughing. after getting a little nauseas, i had to smoke again.

the walk outside seemed much longer than it was before. while outside, i leaned
on a window and stared at the sky. i could hear the music from my room, and
my eyes began to shut. the closed eye visuals were so beautiful. there were
multitudes of stars exploding like fireworks. i felt more and more beautiful with
each toke of my cigarette. i went back to my room, where a friend was chillin
on my bed. i had to use the restroom, but just thinking of going seemed like
such a journey (i live in a dorm and the bathroom is down a short hall). but i
had to go, so i walked the long distance and made my way to the bathroom. the
tile on the floor was breathing and it was very interesting to look at. then, while
i was washing my hands, still feeling pretty from my smoke before, i looked
into the mirror. that's when i learned, **do not look into the mirror!** little red
lines started crawling around my face. i shut my eyes and ran back to my room.

now, i have already lost all concept of time. i closed my eyes, listened to the music and just enjoyed the visuals going through my mind. eyes seemed to be a big visual for me. evil eyes, closed eyes, pink eyes, purple eyes, etc. floated around my mind. it was interesting how the evil eyes didn't scare me at all. i was very fascinated with them.

the song we were listening to kept saying sunlight and i was able to feel the sunlight on my face as the lights in my room seemed to get brighter. **i felt as though i was seeing the world as it should be . . . beautiful!** as the song says "i'm drifting away somehow, by the presence of the morning sun . . . sunlight, there's nothing like your warm embrace. it feels so right shining on your sculptured(?) face." i could feel the warmth and happiness of the world . . . i wanted to share that moment with everyone!

throughout the "trip," i kept thinking of how i would right a trip report because i wanted everyone to feel the way i was feeling. at this point i realized that once my mind had a thought, i was fixated on that thought. i wanted another cigarette because i was still a little nauseas, but i was enjoying the visuals so much (hearts and eyes and rainbow patterns) but a lil cigarette beautifully lit, kept flashing in my mind. my mind was thinking a mile a minute. around 1-ish, i finally went outside for another cigarette and stared at the sky. the stars were dancing in the glow of the moon. it was very enchanting. they all became shooting stars with various colored trails. again, it was another realization that the world and everything in it was beautiful.

i went inside because it was so cold. i was sitting on my bed again, eyes closed, when i started envisioning what a beautiful thing sex was. i kept seeing colorful outlines of a couple merging into one another while they were making love. they were surrounded by various patterns and there was a certain glow around them. they would merge into each other and turn into other patterns like hearts.

then my phone rang. it was one of the other friends who had eaten the shrooms, but had gone into another room to watch a movie. she wanted me the friend that was with me to watch the movie with them. so we decided to go. as i was opening my door, i felt very much like alice in alice in wonderland because i felt so small compared to the door and the walls. then, i had to make a trip to the dreaded bathroom. i sincerely disliked the bathroom, like i developed a fear of the bathroom. my friend waited as she became amazed at the tiled floor. the other two ended up walking in my hall, still very giggly. we went outside to smoke, laughing at the fact that whatever movie they were watching confused them beyond belief. and how we were all a little confused by anything =)

my friend amazed at the floor had gone back to my room. and the other two went back to their movie. i went to get the friend in my room because i wanted all four of us to be together. she decided, however, to go back to her room. when she left, i just wanted to experience my room and being alone before the "trip" ended. then i really lost all track of time. i laid in my bed, and every position i was in was more comfortable than the one before. i felt like my body was twisting and flowing. at one point, my face was nestled in my arm and i was amazed at wow wonderful my skin felt. every now and then i had to open my

eyes to make sure i was still awake. i stared and the walls breathing and patterns on the ceiling. i also became fascinated with my feet (i have a thing with
feet and i don't like them very much)

anyway, my room became my comfort zone and i didn't want to leave. any trip
to the bathroom, i had to convince myself to go. but i came back to my room as
fast as i could, i wrapped myself in my down comforter because it made me
feel secure. cev's still very pleasant.

now i am aware of the time. it is 4am and i'm coming down. i become slightly
agitated with little things. i turn off two star lamps i have hanging from my ceiling. i changed pj's like 3 times. two bracelets i haven't taken off in a long time
become extremely annoying. like i said, i became fixated on things for quite
some time. i took one off with ease. the other was very difficult. it was tight so i
pulled it with as much force as possible. i was using my nails, when i felt one
of them tear. i unbent that nail, and continued to pull off the bracelet. when it
finally came off, i decided to cut my nail, but it when i looked at it again, it was
never even broken. i finally went to bed at around 4:45-5 am.
||edit—i just wanted to add that at this point, all thoughts took on the voice of
seth green's character james st. james from party monster||

again i apologize for this being long.

however, i'm glad i got the chance to see the world as beautiful as everyone
should see it. =)

Example six:
I just heard this song the other day and it pretty much describes a cokeheads
life. I'm not sure if this has been posted before, but for those of you interested,
here's the lyrics . . .

"I need Drugs"—Necro

When I come home from work
I'm fiendin' for an eight-ball
I got crack on my mind
I'm hearing cocaine call
Telling me to beep the dealer to deliver me stuff
Keep it a secret from my wife, cuz she thinks I don't use drugs
There I was, bleeding from my nose and damn
I couldn't breathe but I'm still thinking about the next gram
It's Friday night and I'm not trying to leave my crib doped
I'll kill myself while the dealer's eating Japanese food
I ain't got no pride, I'm buying this shit
I'm lying to myself telling the runner I'm trying to quit
It's all make believe, I pretend that I'm true
When you give me credit, I'm dodging you every chance that I get to

Even if its good, I'll sniff it up in a minute

What does everyone think about it? Pretty accurate, huh?

Performer

Some authors associate certain music performers and groups with drugs and the drug experience. The first example is a response to a post from a fifteen-year-old girl who inquired about the feeling of ecstasy because she wanted to try it:

Ecstasy feels jumbled and out of sync like a radiohead album

The second example is a response to a rather long report given by a member regarding his excursion to a Phish Concert. Talking about the band, the author of the original report makes numerous comments about the band and the songs that they played including "They know how to make you feel high even when you are not." In a response to this long report, the author of a post claims that the music of Phish is like being high:

Honestly I sort of felt tears well up in my eyes there for a second while reading that . . . you very articulately allowed me to re-live the experience.

It's particularly memorable, because during the second show I was rolling so hard, and thinking back on the 20 miles of walking, getting our only bag stolen w/in 5 minutes of entering the grounds, hitch-hiking and the 22 hour roundtrip car ride . . . it just made it all seem so worth it to be in that one place at that very moment in time. I can't put into words what Phish does to me. Their songs sound like they are coming out of me when I'm on a roll. It's like I get high and they get in my head to make music.

Thanks for the report, I really enjoyed it and I'm glad you had a good time. Thanks for the video too.

happy trails

The third example refers to specific performer who is emblematic of the drug culture and describes the themes in the music as descriptive of the drug experience:

I always heard Kobain sing about his heroin experiences and that song lithium describes his trials with that too. I think that all of their music after Nevermind sounds like they're on heroine. I remembered how I felt hearing In Utero when I first tried it and it was almost identical, as if they were able to feel it is make it felt in their music.

In the final example, an author says that the drug experience is described well by Jim Morrison in a particular song and, according to the author, it should because Morrison was a credible user of these drugs. In this example, as with the previous example, the attribution of a drug use identity to the performer produces the performer as a significant other for the individual's self-as-drug user. Caughey (1984) has described extensively how people have come to understand certain experiences based on the demonstrations of these experiences by celebrities:

> Jim Morrison was undoubtedly a drug addict. He describes the feeling.

DJ Style

Authors describe the ways DJs spin music to create certain moods. There is considerable lore on the discussions boards regarding many of the DJs, their music, their personalities, and their drugs of choice. One post describes this kind of personae. The author is explaining the culture of the DJs to someone who claimed that they "just didn't get it":

> Ok look, its more than just music. Djs mix music but they also mix their shit. Its like they each have a signature mix of shit to match their signature style of mix. DJ Aurey smokes blunts and valiums and channels the feeling to his crowd. Anyone who has done the two can listen and know what he's doing. That's how it works, they mixin more than sound.

A second example describes the way an author lists DJs and the drugs he feels are best represented by their mixes:

> Its like each of their methods of mixin resemble different methods of gettin ript. Like DJ Shadow for mixin Primos, Screws for screwin, Bam Bams all about coke, and DJ Armaands shit is the picture of a Roll. Pick your shit, you wanna feel it, you gotta put the right guy on the tables.

Quitting Drugs

There are posts in which authors write about reaching out for help to quit drugs. They also vent or describe the feelings of coming down and withdrawal. Advice from others ranges from how to replicate a high with alternative albeit legal substances to how to take one's mind off of the high. Music is central to all this discourse. In the first example, the author proclaims his determination to quit and the respondent suggests an album to help the quitter replicate the high:

> I think i am going to. Its fucking me up and im smoking it way too much. Like every day in the last month i smoked weed and sometimes twice a day. ANd i cant sleep otherwise.

And i just continually cut classes to smoke. i say that because im in college it doenst matter but then i got caught and have a saturday detention and my parents found out i was cutting
(Response)
try to find a hobby ore something to ceap your mind busy . . . If you miss the high, I find that listening to anything my Incubus really loud can help bring it back a bit, at least it will make you head swim like you were high
my life got a lot better , so i ges so wil yours . . . .
good luck

In the second example, another author responds to the person above trying to quit. The author refers to a song that illustrates the condition of losing motivation that the quitter described. The author of this response post is referring to the song to reinforce the feelings that the quitter is having regarding quitting simply because the song refers to a condition that comes with being high, and it refers to the consequences of that condition:

When I took a two week break from weed, I either had to stay up past 1am to go to sleep, or just use cheap london drugs brand sleeping aids. I'm told melatonin will also help you sleep and doesn't affect REM as much.
Yeah, I'm in college and if I smoke before class the chance of me not going increases a lot. I always think of that song "But then I got high" The only solution is to smoke after class and not before.

The third example is a reply to a member who said she can no longer take the coming down at the end of the night. She fears what withdrawal will feel like. The author of the post refers to a song and the lyrics of that song and says that these lyrics describe (to him) the feeling of coming down:

Ever wanna know what its like to come down? This fits perfectly. Makes me too depressed so I have to get up again. But see how right on they are.

Spineshank—New Disease
Now I can take this, everything I know
Realize that I'm nothing I wanted to be
I can never change anything I've done
Because it's the only thing I have left

Blame myself again for what I didn't do
Never even knew it was coming from me
It changed the way I felt, the worst is yet to come
Because I have gone too far now

so as to not appear as being depressed 100% of the time (which I'm not ;] , I'll quote some Peter Gabriel next.

Discussion and Conclusion

These excerpts from DrugSite posts illustrate the wide range of ways people who use drugs and communicate online talk about the ways music is relevant to making sense of the drug experience. The metaphorical resources offered by the myriad aspects of popular music facilitate both the telling of the visceral, transcendental and other ineffable experiences of selves-as-drug users in everyday life and the subsequent responses from others in tele-copresence that become the virtual looking-glass of the user (Zhao 2005). In terms of a grander metaphor we note that drug experiences are like pop music and its lyrics: they tend to range from the extremes of ecstacy to misery.

The DrugSite community, like other online communities, offers individuals the opportunity to engage in a variety of discourses in order to realize their emergent identities. The freedoms of communication provided by a virtual world where time and space are collapsed and anonymity is all but guaranteed allow individuals to create selves that would otherwise remain suppressed or vulnerable to self-defeating stigmas. The way individuals come to recognize themselves as drug users is critical to their ability to properly manage use, regulate behavior during use, and control their desires if and when they decide to cease use of drugs. Indeed the DrugSite community fosters an ethic of responsible use. A first-time user of the site bragged about engaging in risky and illegal behaviors, such as street racing and vandalism, while being high on cocaine. Another member, speaking for the community responded in kind:

> Dude, nothing about what you are saying is cool!!! We're here to get rid of the stereotypes about what we do. It's people like you who are giving us a bad name. We use drugs, but we also use our heads. If you don't do that then the drugs will use you.

From a clinical perspective, it would be interesting to see to what degree the meanings attached to drug experiences work. Programs for rehabilitation usually fall along three dimensions: the institutional, whereby the whole of the person, body and mind, are surrendered to some institutional authority; the group-support, where the whole of the individual, body and mind, is surrendered to some transcendental authority (to God or to the group); and inner-driven, where the individual turns inward and re-constructs the self. The first two of these dimensions often rests on acknowledgements of powerlessness that can, in themselves, have unintended consequences for the individual.

The latter of the three is often the most effective for long-term change in that it creates change through the determined actions of an autonomous agent. The existential perspective in sociology (Kotarba and Johnson 2002) tells us that there is great personal as well as social comfort in simply having meaning for a troubling experience or phenomenon. Considering this in light of the findings in this chapter would suggest that that the freedoms to construct and make sense of a self-as-drug user, independent of other social labels or stigma, could reduce the anxiety resulting from suppressed identities and promote a more autonomous identity; one that will be more efficacious in future attempts at cessation. While this is only conjecture, it points to several possible designs. In particular, a follow-up study of drug users who invoke musical resources to help them quit drugs in effort to assess their perceptions of the advice they received on Drug-Site, specifically in regards to music, would be fascinating for both the basis of drug research and clinical application perspectives.

References

Altheide, David. "Identity and the Definition of the Situation in a Mass-Mediated Context" *Symbolic Interaction*, 23, no. 1 (2000.):1-27.

Adams, Paul. *The Boundless Self.* Syracuse, NY: Syracuse University Press, 2003.

Becker, Howard. "Becoming a Marijuana User." *American Journal of Sociology.* 59, no. 3 (1953.):235-242

———. *Outsiders: Studies in the Sociology of Deviance.* Chicago: The Free Press, 1963.

Caughey, John. *Imaginary Social Worlds.* Lincoln, NE: University of Nebraska Press, 1984.

Charmaz, Kathy. "Grounded Theory: Objectivist and Constructivist Methods." In *Handbook of Qualitative Research* edited by Norman K. Denzin and Yvonna S. Lincoln (Second ed.). Thousand Oaks, CA: Sage, 2000.

Denzin, Norman. *The Alcoholic Self.* Newbury Park, CA: Sage, 1987.

Glaser, Barney and Anselm Strauss. *The Discovery of Grounded Theory.* Chicago: Aldine, 1967.

Goffman, Erving. *Stigma: Notes on the Management of Spoiled Identity.* New York: Simon and Schuster, 1963.

Holstein, James and Jaber Gubrium. *The Self We Live By: Narrative Identity in a Postmodern World.* New York: Oxford University Press, 2000.

Kotarba, Joseph A. "Baby Boomer Rock 'n' Roll fans and the Becoming of Self." In Joseph A. Kotarba and John M. Johnson (eds.) 2002.

———. "The Positive Functions of Rock 'n' roll Music for Children and their Parents." in *Troubling Children: Studies of Children and Social Problems* edited by Joel Best (ed.). New York: Aldine, 1994.

Kotarba, Joseph A. and John M. Johnson (eds.). *Postmodern Existential Sociology.* Walnut Creek, CA: Alta Mira, 2002.

Lakoff, George and Mark Johnson. *Metaphors We Live By.* Chicago: University of Chicago Press, 1980.

Surratt, Carla. Netlife: Internet Citizens and their Communities. New York: Nova Science, 1998.

Turkle, Shirley. Life on Screen: Identity in the Age of the Internet. New York: Simon and Schuster, 1995.

Zhao, Shanyang. "The Digital Self: Through the Looking Glass of Telecopresent Others." *Symbolic Interaction*, 28, no. 3 (2005):387-405.

Chapter Thirteen

A Review of Internet Studies in this Volume, an Examination of Root Causes of Drug Abuse from a Societal Point of View, and Some Possible Solutions

Edward Murguía and Ann Lessem

This volume is intended to solicit new and pertinent information about a relatively unexplored segment of psychoactive substances and their use, namely, club drug use and the role of the Internet in transferring and mediating information about drug use across settings. To that end, researchers in this volume have entered the world of drug users and ravers, have traversed cyber space domains from harm reduction to no-tolerance, have examined concepts of freedom and community, have stretched ethnographic methodologies, have imposed old theories on new settings and new theories on old settings, and generally have addressed controversial issues in need of study and debate.

A Review of the Studies in this Volume

Gatson begins the journey by posing the question, "what are the connections between online activities of drug users (or potential users) and offline drug use?" Concerns about this issue arose because of the perception that Internet communications are an uncontrolled/uncontrollable means to influence individuals and communities. Seeking information via the Internet is also quite different than the traditional offline scenario of information seeking. Using traditional, offline methods, the individual seeking information can typically be identified and has access to a relatively small number of information sources. However, online, this individual is more or less anonymous and can link to "virtually" unlimited sources of information and/or persons.

In order to answer her research question, Gatson focuses her investigation on close-knit Internet communities centered on drugs and music/raves. She then adapts traditional ethnographic methods to the practice of cyber-ethnography. Using these methods, she finds an amorphous subculture that is interlinked online and offline. She also finds the merging of both transgressive and "progressive" (in the sense of the development of new and adaptive cultural practices) concepts and practices. Technology in many forms is of tantamount importance and is used in the creation of "techno" music and video production; is incorporated into media and communication; is used to produce the drugs of choice (MDMA, GHB, etc.); and is the basis for an informal and anti-corporate "do-it-yourself and share" value system connected to these technologies and chemical practices.

Gatson concludes that it is crucial to understand the intensity, purpose, and effectiveness of online interactions in order to understand the people who make and disseminate them. Therefore, ethnographic methodologies used as an extended case method are critical to understanding issues of shape, density, content, and contact for people who are engaged in this complex scene.

Adjusting to a sharper focus, Murguía, Crispino, and Tackett-Gibson "get into the heads" of drug users as they examine reasons for use, deterrence, and acceptance of risk. Murguía asks the basic question, "Why do people use drugs?" He starts with five theoretical models from the deviance literature used to explain drug use and compares them to a conversation drug users had about their motivation to take drugs. The theoretical models were:

1. Gottfredson and Hirschi's "self control theory" that is grounded on bio-psychological elements of self-control and contends that the basis of this self-control is the ability to recognize the negative consequences of deviant behavior;

2. Brook's psycho-sociological "family interaction theory," that posits an association between low drug use and an affectionate, non-conflictual relationship with parents during childhood;

3. Hirschi's "social control theory" that takes a sociological approach in contending that the individuals most likely to be delinquent are those not attached to normative others and who therefore lack hope for achievement within conventional society and who do not have conventional moral beliefs;

4. Oetting and Beauvais' "peer cluster theory" that gives central place to the influence of peers to teach, sustain, and sometimes to increase drug use, al-

though the first three models more fundamentally explain the causes of drug use than do peer influence; and

5. Sykes and Matza's "techniques of neutralization theory" that delineates justifications used to validate behavior, including the denial of responsibility, the denial of injury, the denial of victim, the condemnation of condemners, and the appeal to higher loyalties.

With these theories as a backdrop, Murguía uses qualitative constant-comparative methodology to analyze an online conversation where drug users discuss why they used drugs. This analysis results in six basic answers to the question of "why" and two additional crosscutting considerations.

The answers drug users gave to the question of why they used drugs were:

1. To satisfy an addiction;
2. For self-medication;
3. To avoid problematic reality;
4. For happiness and pleasure;
5. For friendship; and
6. For insight and inspiration.

And the two crosscutting factors that also emerged from their conversation were:

1. Different drugs have different effects and are used for different purposes; and
2. For many users, drug use changes over time.

While some might interpret the causal deviance theories and the conversation of the drug users as contradictory, Murguía believes that they are complementary aspects of a complex phenomenon. He hypothesizes that the causal deviance theories describe distal factors in drug use and emphasize the psychological/social/physiological discomfort developed early in life as the fundamental basis of drug use. On the other hand, the conversations of the drug users are consistent with proximal factors of drug use and emphasize more immediate concerns and issues. In other words, the factors from the theoretical models (being distal) point to causes of discomfort while the factors from the phenomenological analysis of conversation (being proximal) point to relief from that discomfort and/or life enhancement that may (or may not) come, with the caveat that drug use can lead to addiction, self-medication, and avoidance of reality in the future.

Crispino then asks about the intent of drug policy and the deterrence of drug use. She explains that U.S. drug policy is primarily a combination of rational choice theory and deterrence theory in which the government attempts to deter drug use through formal and informal sanctions that are stringent enough to overcome the apparent gain one would experience from using a drug. She argues that these sanctions do not play as significant a role in shaping drug use decisions as traditionally thought. By analyzing discourse both from parents attempting to deter their children from using drugs and from drug users themselves, she concludes that drug users and their families are more concerned with the intrinsic effects of drug use than with sanctions. She dubs this phenomenon "causal-empirical deterrence."

Crispino cautions us that, in applying rational choice theory and deterrence theory to drug use, one first must make the assumption that drug use is deviant. However, if drug use is not considered deviant, then the deterrence methods employed by the government will be less than effective. Consistent with several of Murguía's findings, Crispino finds that drug users themselves often see their behavior only as pleasurable. Additional arguments against drug users as being deviant stem from the inconsistent ways in which drug laws are applied among different social classes, between blacks and whites, and between men and women. For example, middle class white drug users are unlikely to label themselves as deviant because they are able to keep their drug use secret. On the other hand, lower class black drug users are more likely to become publicly known for their drug use, are more likely to be publicly labeled as a deviant, and are more likely to suffer greater consequences for their drug use.

Crispino finds that a reasonable way to go about establishing whether drug use is generally or subculturally considered deviant is to read what people write about this topic online where they are anonymous and where there are relatively few consequences attached to what they might express. She examines two very different websites; one that bills itself as an anti-drug resource for parents who want to deter their children from using drugs and one that prides itself on its harm reduction stance and believes that it is more important to provide information that makes use as safe as possible. What she discovers from both sites is that rational choice and deterrence theories are correct in assuming that users weigh the possible gain from drug use against the possible negative effects. However, these negative effects are considered in terms of causal-empirical deterrence intrinsic in the substance itself and not in terms of formal and informal sanctions. She also discovers that any negative effects of illegal drug use presented dishonestly are discounted and have little effectiveness as deterrents to illegal drug use.

Tackett-Gibson expands upon the theory behind drug use as she uses online discourse to study whether online behaviors of individuals seeking information from a harm reduction Internet site are consistent with postmodern theories of "risk society." She examines voluntary risk behaviors, how individuals who engage in these behaviors confront health risk, and how online discussions help to define and mediate use, risk, and consequences. Ironically, the primary risk discussed by Tackett-Gibson is not a risk associated with drug use, but the risk associated with the ease of use and proliferation of online information about drugs. Internet use can be uncertain and unpredictable and it is often difficult to discern the legitimacy of the information obtained.

Tackett-Gibson obtains her data from transcripts of online threaded discussions that took place at a large, popular drug use information site that is known for its harm reduction stance. Her findings highlight the self-reflexive nature of voluntary risk-taking, that is, risk-taking behaviors are assessed, analyzed, and described in terms of benefits and harms and it is each individual's personal responsibility to assess and manage harm and pleasure. However, she points out that risks are only manageable with adequate information and knowledge. Because "knowledge" about drugs is often contested to the extent that the identification of the "truth" is problematic and political, the participants at the site value the knowledge and experience of other site members and reject the accuracy of other publicly accessible sources of knowledge such as school education programs and government efforts to reduce drug use.

Tackett-Gibson's analysis of online drug discourse shows several theoretical characteristics of a "risk society." These are:

1. High-tech methods are used to assess the risks of high-tech artifacts;
2. Knowledge is mediated through social processes that are a product of virtual interactions;
3. In a harm reduction community, risks are second-hand rather than experienced;
4. Risks are defined through a process of interaction reliant on interpretation of scientific data and personal accounts of use;
5. Risks are unseen and left to be experienced at a future time—when it is too late to mediate risk consequences; and
6. Risks are illusive and undefined.

Tackett-Gibson in her chapter on prescription stimulants online indicates that prescription drug abuse is of major concern, and that the rate of use of prescription drugs for non-medical reasons is higher than that of cocaine, hallucinogens, or ecstasy. The advantages of using prescription drugs for non-medical reasons as opposed to using "street" drugs are clear; prescription drugs tend to be "clean," that is, unadulterated, and they can be obtained easily either legally and at a low cost with a prescription subsidized by medical insurance, or from others with a prescription. Her focus is on a subset of prescription drugs, namely, stimulants such as methylphenidate (Ritalin), amphetamine (Adderall), dextroamphetamine (Dexedrine) and methamphetamine (Desoxyn), and she indicates that information on Ritalin alone is vast—ten million active websites contain information on Ritalin. She finds that college students are using Ritalin intranasally for performance enhancement. Interestingly, in the universe of drug users, prescription drug users tend to have relatively lower status than "street" drug users.

Among prescription drug users themselves, lines are often blurred among use for medical purposes, recreational purposes, or because users are addicted. Finally, Tackett-Gibson indicates that the fact that a drug is obtained by prescription may make individuals forget about the potential dangers of that prescribed drug.

Kotarba (both as a single author and as a co-author with Halbert) and Lessem move from the realm of general drug use to the realm of drugs and music. Kotarba uses a harm reduction Internet site to explore both the relationship between specific drugs and specific types of music and also the way that drug users use music to interpret their drug experiences. Lessem then turns the tables and examines drug discourse on a music/rave Internet site.

Kotarba begins with a summary of the history of the relationship of popular music and illegal drug use. He then explains that "members of the scene" are likely to engage in a phenomenon he calls *drugmusictalk*. The primary research question for Kotarba's first study is "what is the structure and function of *drugmusictalk* in the context of the Internet? The theoretical framework for the study is symbolic interaction, i.e., social life is constructed through various forms of interaction and people interact with each other pragmatically to arrive at solutions for shared problems, most often experienced in concrete situations. Kotarba conducts a content analysis of forums on a harm reduction Internet site and then categorizes elements of *drugmusictalk* inductively, following the logic of grounded theory. By doing so, he finds that participants at the Internet site use *drugmusictalk* in the following ways:

1. To share aesthetic drug-music experiences;
2. To discuss the fit between particular drugs and styles of music;
3. To share affective aspects (e.g., sexual) of drug-music experiences;
4. To discuss economic and/or ideological aspects of drug-music experiences; and
5. To discussing personal tastes in music.

He concludes that the Internet site appears to be a desirable and useful venue for talking about one's music, dance, and drug experiences holistically, as they occur in everyday life. From the participants' perspective, the Internet site provides them the opportunity to share experiences with others very much like them.

In the Halbert and Kotarba study, they ask the question, "What are the various ways that participants who are communicating with each other via a harm reduction website use musical phenomena to make sense of their drug experience?" To find the answer, they conduct a content analysis of messages over a one-year period and use a grounded theory style to draw conclusions.

Halbert and Kotarba find that music-related posts appear consistently throughout the harm reduction site posts and generally fall within six categories:

1. Musical style used metaphorically to describe drug experiences;
2. Specific songs used metaphorically to describe drug experiences;
3. Lyrics used to describe drug experiences;
4. Performers associated with drug experiences;
5. DJ styles associated with drug experiences; and
6. Quitting drug use.

Halbert and Kotarba conclude that drug experiences are like pop music and its lyrics: they tend to range from the extremes of ecstasy to misery. They also find a consistency with an existential perspective in sociology in the way individuals using music as a means to talk about drugs find social comfort in finding meaning for an experience or phenomenon.

Lessem then looks at the relationships among music, drugs, and online discourse from "the other side of the mirror" as she examines a free-flowing conversation on a rave-specific Internet site to determine if drug-related conversations occur and, if so, how they impact drug use. In the course of conducting her ethnographic analysis she finds multiple communication anomalies that include:

1. Conversations that were both synchronous and asynchronous, i.e., some individuals were online at the same time communicating with each other and other individuals came online at different times and interjected comments into a conversation after that conversation had concluded;
2. Conversations that were both inclusive and exclusive, i.e., the conversations were open to anyone, but used such specific rave-related slang that only members of the group could understand the discourse;
3. Conversations that ranged in scope from profound items that reflected on participants' lives and their place in the world, to conversations that were explicitly sexually profane;
4. Conversations that ranged in topic from the fantastic (e.g., bizarre fictional stories developed on the fly), to the mundane, to the surprising (e.g., investment strategies);
5. Conversations that used pictures instead of words, and pseudo conversations that used words as an art form (e.g., repeating the same word over and over and over and over).

Interspersed throughout the anomalous conversations of the ravers were numerous references to drugs. During these discourses the ravers talked about drugs in four different ways:

1. They used drug-related slang as a conversational means to identify themselves as members of the group;
2. They used humorous references to drugs as a means of entertainment;
3. They used drugs as a prop and/or an event in their fantasy stories; and
4. They talked seriously about drugs and drug use—often about the negative impact of taking drugs and of having drugs at raves.

Lessem concludes that in order to more thoroughly understand raves and their relationship to drugs, we must more thoroughly understand the heterogeneity of ravers. Contrary to the assumption of many anti-drug organizations, she also discovers that the drug-related conversations of the ravers are not a peril that must be curtailed in order to protect youth. Instead, they are a portal through which the lives of youth who attend raves better can be understood.

Gatson then pulls the focus back to harm reduction as she examines the history of the movement and its expansion online. Her premise is that the move-

ment of harm reduction to online venues has the potential to open up the definition from the micro level and public policy/medical issues to the macro level,
that is, to the political and legal issues level. She reminds us that moral panics
are not new and that they recur around media, technology, and the body/mind
generally. While the language of these moral panics often focus on the young
and innocent, the risk of harm to the minds and bodies of the young that the
campaigns articulate are not primarily about those individual young people. Instead, the panic is about the reproduction of particular mores—the group
boundaries of sanctioned behaviors and ideologies. Thus, linking Internet use
and drug use is simply an example of moral panics that can lead to extreme terror because the symbolism of drug use attaches itself to current popular fears
and can become so intense that it spills over into other areas of concern.

Gatson concludes that harm reduction seems to come from a place of cultural pluralism and liberalism and focuses on the potential and actual harm to
particular individuals rather than on a sense of harm to the boundaries of the
group, or between groups within a society. However, it is harm to our morals
that is the underlying structure of the debate about the viability of the harm reduction movement. She then cautions that no moral panic campaign has ever
succeeded in stamping out whatever its targeted behaviors have been.

In her final study, Gatson examines spaces where youth might be exposed to
communication about illegal and/or dangerous drug practices. The organizing
question for her analysis involves understanding both the Internet environment
and the issue of drug use as situations of relative anomie as she asks, "are these
folks innovating or rebelling—or are they possibly even conforming to the democratic tradition?" Gatson goes on to point out that any society has some
more-or-less stable structures and some more-or-less boundary pushing behavior
aimed at (or affecting) those structures. How people go about living within those
structures involves taking them for granted as well as articulating explicit adherence and/or challenges to them. Thus, folkways (norms) and mores (laws) must
be seen to exist in more of a state of flux than in the analytically distinctive
sense that understanding them as ideal types affords.

Gatson calls the bulk of the individuals who make up the participants on the
websites she assessed "politically engaged deviants," or "counterpublics." They
challenge the status quo in terms of both norms and laws, but they do it in common cultural discourse, emphasizing their central place in the arena of citizenship. She believes that such politically engaged persons are often at the forefront
of social change and that they also focus our attention upon the symbolic
boundaries of our current consensus about what is right and what is wrong. The
field, or scene, of drug-using discourse thus shows us that society may be understood as an argument. Discourse within this scene often acknowledges the risky
nature of drugs use as it argues for its potential to expand the mind, the self, and
society.

When those engaged in deviant behavior resist the label of deviant with a response that is articulated as non-deviant, indeed crucial to the normal process of
politics and making law, they highlight the complexity of labeling itself, as well

as conflicts over situation, audience, or interest/reference group. When those engaged in online discourse about drugs were asked, "if you could sit down with us and tell us anything that you think we should know about recreational drug use, music and raves, online communities, or the government's policies toward drugs, what would you say?" they argued for free thought, free speech, free information, and the free action of the harm reduction discourse.

Gatson concludes by reminding us that there are many policy implications, at several levels of government, inherent in recognizing that the nexus under investigation in this volume are embedded in ongoing discursive (legal/political) traditions. Her list of examples include:

1. Health policy—where there could be more official and open collaboration between the National Institute for Drug Abuse/National Institutes of Health and various arms of the harm reduction movement, and where there is a need to recognize that the promulgation of exaggerated hysterical ad campaigns and easily contestable data do little to stop harmful drug use and may do much to increase cynicism and distrust of health care authority figures;

2. Local policy—where local approaches to dealing with the (perceived) problem of extant and/or increased childhood and adolescent drug use and abuse could be developed to take advantage of extant models of Internet discussion groups and could focus on providing accurate information to youth;

3. State and federal policy—where there is conflict between states, and/or between states and the federal government, over drug policies and where lawmakers and citizens could be supported in forming formal Internet discussion groups;

4. International policy—where all governments could take advantage of the multiuser communication and community functions of the Internet to listen to the citizenry at large.

Finally, Gatson asks the hard question we all should consider in our examinations of illicit substance use and the Internet—"are the risks inherent in drug use worth tolerating in order to ensure the values inherent in the also risky business of free speech?"

An Examination of Root Causes of Drug Abuse from a Societal Point of View, and Some Possible Solutions

Although drug use is universal, we believe that there are root causes of drug use that can be attributed to the socioeconomic structure of the United States. What causes individuals in the United States in particular to take mood-enhancing substances? An understanding of root causes of drug use in the United States from a societal perspective will enable us to begin to develop antidotes to excessive and harmful drug use.

We believe that a fundamental cause of extreme drug use can be traced to the philosophy held by individuals, derived both from their family and, more broadly, from their society and their culture, composed of the often unstated but fundamental beliefs that govern how they see and judge events in their lives. In his influential book, *American Society: A Sociological Interpretation*, Williams (1960) states concerning American culture, "First, American culture is marked by a central stress upon personal achievement, especially secular occupational achievement" (p.417). Additionally, personal achievement should be based on hard work and self-sacrifice, and must be gained honestly.

The value of personal achievement has been and continues to be expressed in American society in numerous ways. The Horatio Alger "rags to riches" books, for example, clearly express this theme, as does the high regard in which, for example, Benjamin Franklin, one of the most popular of the founding fathers of the United States, is held. Franklin arrived in Philadelphia penniless as a young man and became one of the wealthiest Americans of his time; many of the sayings in his "Poor Richard's Almanac," such as "Early to bed, early to rise, makes a man healthy, wealthy and wise," contain instructions as to how to attain a high degree of personal achievement. The greatest of all American heroes, Abraham Lincoln, personifies secular occupational achievement because he was born in a log cabin, and yet, he became President of the United States.

At the other extreme, an archetypically tragic figure in American literature is Willy Loman in Arthur Miller's insightful play, "Death of a Salesman." Loman believed himself to be a failure and committed suicide, even though he made a decent living as a salesman and was able to support his family. He considered himself a failure for not being able to rise above what he thought to be his own unacceptably low level of achievement.

A clear vision of the value of personal achievement in American society can be gained through an analysis of sport in America. In sport in America, competition is structured so that there must be a winner and a loser of every game. Situations are created so that eventually, all but one must lose; only one can be considered "champion" in this zero-sum game structure. Since self-esteem in the United States so greatly is based on achievement, and since, at least in theory, all participants have an equal chance to win, loss becomes a blow and can result in a loss of self-esteem. To ameliorate some of the effect of this philosophical stance prevalent in the United States as to the importance of achievement, and because American culture teaches that anything less than being the "best" is failure, and because, given the structure of American society, almost all of us will "lose," and feel ourselves to be failures at some point, we believe that mood-enhancing substances are taken by numerous individuals in the United States to cope with the pain of loss of self-esteem, or in order to enhance performance and thus to "win." In other words, Americans self-medicate either to feel better after almost certain failure, certain given the structure of competition and the inevitability of loss, or they take drugs in order to "win" so as to postpone the self-esteem damaging effects of failure.

In film, Marlon Brando's portrayal of an ex-prizefighter in "On the Waterfront" when he exclaims to his brother who caused him to lose, "I could have been somebody. I could have been a contender," is another illustration of this feeling of failure endemic to life in the United States. Failure at a particular time and in a particular setting frequently is exaggerated into the idea that one is "a total failure." This thought in particular is a psychic distortion that does enormous damage to individuals, leads to psychic pain, and results in self-medication through excessive drug use.

Lipset's (1967) study of American society in *The First New Nation* also provides insight into the causes of drug use in the United States. He believes that the two fundamental values in American society are achievement (in this he is in agreement with Williams) and egalitarianism. The belief in egalitarianism from the point of view of a belief that everyone has the potential of achieving "greatness" of some kind in the United States sets us up for distress. Failure, which as we have said is universal in our society, is seen as a personal failure. We do not see ourselves as locked into a class system that predestines many of us to an existence of subservience, which would result in our blaming social forces for our lack of success. Instead, we see ourselves as having the potential of achieving everything, and since, of course, this is impossible, and we do fail, we suffer from our own sense of failure, of not having done as much as we could have done. Ironically, then, drug use is the downside of the American belief in the potential for universal upward mobility.

What then, is a possible antidote to the drug problem in the United States? We must remember that while other nations supply us with drugs, it is we in the United States who actually buy and use illegal mood enhancing substances. The drug problem in the United States is our problem. We believe that the founders and explicators of cognitive behavioral psychology have a large part of the answer. David Burns ([1980] 1999) whose work is based on the work of the founders of cognitive behavioral psychology, Albert Ellis (1975; [1988] 1990)) and Aaron Beck (1967; 1991; 1997), states that there is no such thing as global individual failure as in, "I am a failure" or "I am worthless" or "I am no good." It is thoughts such as these that indicate a loss of self-esteem, which lead to extreme discomfort, and that lead to self-medication with drugs. What follows is a brief summary of some of the main tenets of cognitive behavioral psychology:

1. Move away from the idea that you **must** have some job or promotion, some object, or the affection of some person. Substitute "I would prefer to have" for "I must have," and realize that you can still be happy and live a full and satisfying life, though perhaps not as happy, without the object that you desire. You can negotiate with others to obtain what you want, but if you do not obtain it, you will still be fine. This is a central point in many of the works of Albert Ellis. (See for example Ellis (1975; [1988] 1990).

2. Move away from "My work is my worth" to unconditional self-love and unconditional self-esteem. This idea is best explicated in Burns ([1980] 1999:327-51). In other words, you don't have to achieve anything to be a

worthy human being. Achievement is nice, but it has nothing to do with being a worthwhile human being.

3. There is no such thing as "I am a (total) failure." This is an exaggeration. All of us at times do some things well, and other things not so well. Labeling oneself as a "failure" is destructive. (See for example, Burns ([1980] 1999:90-91)).

4. Move away from "Anything worth doing is worth doing well" to "Anything worth doing is worth doing badly." In other words, do things for themselves, for the joy of doing the task itself, not for fame, glory, and applause from others. If you enjoy it, then do it and do not drive yourself to be the "greatest" ever to do it. It is a good idea to "dare to be average" in the sense that if you try to be perfect, you drive yourself constantly and you never derive any satisfaction from anything that you are doing because ultimately, nothing can be perfect. There is always another level and if you constantly try for it, it can leave you exhausted and doubting yourself. (Burns ([1980] 1999:352-80)).

A second antidote to the destructive effects of uncontrolled ambition in the United States resides in Buddhist philosophy, becoming increasingly popular among some segments of the American population. The "Four Noble Truths" of Buddhist philosophy are as follows:

1. Life consists of suffering and disappointment
2. Desire for pleasure, power, and existence cause suffering
3. One needs to stop desiring to stop suffering and disappointment
4. One stops desiring through the "Noble Eightfold Path," namely, right views, right intention, right speech, right action, right livelihood, right effort, right awareness, and right consciousness (*New Encyclopaedia Britannica* 1993:602-3).

Again note that the emphasis is on a lessening of desire, and this places this religious philosophy in line with cognitive behavioral psychology. Two of the most influential Buddhists in the United States at this time are the Tibetan Buddhist, Tenzin Gyatso, fourteenth Dalai Lama (see for example, 1998), and the Vietnamese Buddhist, Thich Nhat Hanh (for example, [1975] 1987; [1995] 1997.).

Drug Policy in the United States

Concerning drug policy in the United States, we must remember that not all mood-enhancing substances have the same effect on the human body, and that not all mood-enhancing substances are used in the same contexts and with the same purpose in mind. On the most positive side concerning drugs, in this case alcohol, we present a quotation from Robert Mondavi (1998), credited with giving prominence to the California wine industry and, in particular, to the wine industry in the Napa Valley of California. Mondavi's parents were immigrants to the United States from central Italy.

> Meals were a sacred time for us, the gathering of the family, and wine always enjoyed a prominent place on our table . . . . I was trying to plant deep into the soil of our young country the same values, traditions, and daily pleasures that my mother and father had brought with them from the hills and valleys of central Italy: good food, good wine, and love of family . . . . (P. 19-20). Wine was part of my experience as early as I can remember. At lunch and dinner, my mother of father would routinely flavor my water with a dollop of wine. I didn't like water alone, but I loved it that way. So I grew up thinking of wine as liquid food. It tasted good, and everyone though even then that it was healthy—good for the circulation and a general tonic for the system. (P.113).

Mondavi, born in 1913, is ninety-two years old as this is written.

On the other hand, we need to recognize the harmful effects of some drugs in some contexts. These quotes from the diary of a fifteen-year-old girl (Anonymous 1971) with serious socio-psychological problems, daughter of a college professor, in the midst of recovering from a drug overdose of LSD. She died later in the year of a subsequent drug overdose.

> I have tried to piece the whole thing together but I can't. The nurses and doctors keep telling me I will feel better, but I still can't get straight. I can't close my eyes because the worms are still crawling on me. They are eating me. They are crawling through my nose and gnawing in my mouth and oh God . . . . (P.122). The worms are eating away my female parts first. They have almost entirely eaten away my vagina and my breasts and now they are working on my mouth and throat. I wish the doctors and nurses would let my soul die, but they are still experimenting with trying to reunite the body and the spirit. (P.123).

The U.S. government needs to continue to treat different kinds of mood-enhancing substances differently. For example, the caffeine in coffee and the alcohol in wine and beer are and need to be treated differently than cocaine, methamphetamine, and heroin. One of the most problematic drugs at this time in terms of social policy in the United States is cannabis. Many believe that the penalties for cannabis (marijuana) possession in the United States are not in line with the harm actually done by the drug. It is of interest to look at how the Netherlands treats the issue of the criminalization of cannabis (Wikipedia 2006). In the Netherlands, drug use in general is considered to be a public health issue rather than a criminal issue, and a distinction is made between "hard drugs" such as cocaine, heroin, and ecstasy, and "soft drugs" such as psilocybin mushrooms and cannabis. Soft drugs are considered to be psychologically addictive whereas hard drugs are considered to be physically addictive and are controlled more strictly than soft drugs. Neither the possession nor the production of small quantities of soft drugs are prosecuted, although technically, soft drugs are still considered controlled substances, and possession is still a misdemeanor. Since there seems to be no diminution either in the availability or the use of cannabis in the

United States in this time (see, for example, Community Epidemiology Work Group 2003: 37-42), and because the resources spent by our criminal justice system related to cannabis use are very large, it may be wise to revisit this issue, although the political climate may not allow a revisiting of the decriminalization of cannabis at this time.

Finally, the 100-to-1 crack cocaine/powder cocaine quantity ratio for federal mandatory minimum sentences for trafficking and for simple possession is extreme. Crack cocaine is made from powder cocaine at almost a 1-to-1 ratio (1 gram of powder cocaine makes .89 grams of crack cocaine). The major difference between the two substances seems to be a racial disparity between users of crack cocaine, largely African American and Hispanic, and powder cocaine, the majority of users being white. In 1995, the United States Sentencing Commission, charged by Congress with setting federal sentencing guidelines, recommended a revision to this ratio, but Congress refused to make any change (The Sentencing Project, 1997). In 2006, the American Civil Liberties Union joined several distinguished law professors in a coalition brief arguing against the 100-to-1 crack/powder cocaine ratio. In 2000, 94% of those convicted in trafficking crack cocaine were African American or Hispanic and only 6% were white (American Civil Liberties Union 2006).

In sum, we have reviewed the chapters in this book, have outlined societal structural root causes of drug use in the United States, and have considered some possible antidotes to drug abuse in the U.S. Finally, we have made some policy recommendations that we believe would benefit American society concerning both the legal treatment of cannabis and the sentencing guidelines concerning crack cocaine. It is our hope that a thoughtful consideration of these issues will move us toward a more enlightened and a more just society.

References

American Civil Liberties Union. 2006. *ACLU and Sentencing Experts Urge Federal Court to Uphold Judges' Right to Reject 100-1 Crack-Powder Ratio.* American Civil Liberties Union. Retrieved April 3, 2006 (http://aclu.org/drugpolicy/sentencing/23568prs20060120.html).

Anonymous. *Go Ask Alice.* Englewood Cliffs, NJ: Prentice-Hall, 1971.

Beck, Aaron T. *Depression: Clinical, Experimental, and Theoretical Aspects.* New York: Harper and Row, 1967.

———. "Cognitive Therapy: A 30-Year Retrospective." *American Psychologist* 46 (1991):368-75.

———. "Cognitive Therapy: Reflections." In *The Evolution of Psychotherapy: The Third Conference,* edited by J.K. Zeig. New York: Bruner/Mazel, 1997.

Burns, David D. *Feeling Good: The New Mood Therapy.* New York: Avon, [1980] 1999.

Community Epidemiology Work Group, National Institute on Drug Abuse. 2003. *Epidemiologic Trends in Drug Abuse, Highlights and Executive Summary.* Bethesda, MD: Department of Heal and Human Services.

Dalai Lama and Howard C. Cutler. *The Art of Happiness: A Handbook for Living.* New York: Riverhead Books, 1998.

Ellis, Albert. *How to Stubbornly Refuse to Make Yourself Miserable About Anything—Yes, Anything.* New York: First Carol Publishing Group, [1988] 1990.

Ellis, Albert and R. A. Harper. *The New Guide to Rational Thinking.* North Hollywood, CA: Wilshire Books, 1975.

Lipset, Seymour Martin. *The First New Nation.* Garden City, New York: Anchor, 1967.

Mondavi, Robert. *Harvests of Joy: My Passion for Excellence.* New York: Harcourt Brace, 1998.

The New Encyclopaedia Britannica. Fifteenth Edition. "Buddhism." Chicago: Encyclopaedia Britannica, pp. 602-3, 1993.

Nhat Hanh, Thich. *The Miracle of Mindfulness: An Introduction to the Practice of Meditation.* Boston: Beacon Press, [1975] 1987.

———. *Living Buddha, Living Christ.* New York: Riverhead Books, 1995.

The Sentencing Project. 1997. "Crack Cocaine Sentencing Policy: Unjustified and Unreasonable." Retrieved April 3, 2006 <http://www.sentencingproject.org/pdfs/1003.pdf>.

Wikipedia. Drug Policy of the Netherlands. <http://en.wikipedia.org/wiki/Drug-policy-of-the-netherlands> Retrieved November 3, 2006.

Williams, Robin M., Jr. *American Society: A Sociological Interpretation.* Second ed. New York: Alfred A. Knopf, 1960.

About the Authors

Azzurra Crispino received her M.A. in philosophy from Texas A&M University in 2006, and her A.B., also in philosophy, from Ripon College in 2002. As an undergraduate student, she was involved in research on Computer Mediated communication of the Iowa Student Computer Association Bulletin Board System (ISCA BBS). Her current research focuses on moral compromise with integrity in public policy, with a focus on drug policy solutions.

Sarah N. Gatson, Ph.D. is an assistant professor of sociology at Texas A&M University. She received her doctorate from Northwestern University in 1999. In addition to her work on CMC in rave and drug-using subcultures, she has published several pieces on Internet community, including "Choosing Community: Rejecting Anonymity in Cyberspace," *Research in Community Sociology*, volume 10, August 2000, 105-137; "www.Buffy.Com: Cliques, Boundaries, and Hierarchies in an Internet Community," in *Fighting the Forces: What's at Stake in Buffy the Vampire Slayer*, edited by Rhonda Wilcox and David Lavery, Rowman and Littlefield, 2002, 239-49; and "'Natives' Practicing and Inscribing Community: Ethnography Online," forthcoming, *Qualitative Research*, 4 (2) 2004: 179-200, and the forthcoming book, *Interpersonal Culture on the Internet—Television, the Internet, and the Making of a Community*, Lewiston, NY: The Edwin Mellen Press: Studies in Sociology Series, no. 40, all with Ms. Amanda Zweerink. Her other research interests include gender and U.S. labor policy; the social construction of race, gender, class, community, and citizenship at the nexus of law and culture; and the construction of multiracial identity.

Shawn Halbert is a doctoral student in sociology at the University of California at Los Angeles. His study, "Rhythm and Reminiscence: Popular Music, Nostal

gia and the Storied Self," won the Best Senior Thesis Award from the Honors
College at the University of Houston.

Joe Kotarba, Ph.D. is a professor of sociology at the University of Houston. He
is a graduate of the doctoral program in sociology at the University of California, San Diego. He has a research interest in the areas of sociology of culture,
everyday life, and deviant behavior. He is the recent co-editor John M. Johnson
of *Postmodern Existential Sociology*, published by Walnut Creek, CA: Alta
Mira (May, 2002). He has served as a reviewer, editor, and advisory board
member for various academic journals such as *Sociological Quarterly*, *Symbolic
Interaction*, *Journal of Health and Social Behavior*, *Symbolic Interaction*, *Journal of Contemporary Ethnography*, and *Youth and Society*. He has also served as
a principal investigator on many research grants including: "Health Care Activities Among the Homeless" (funded by the National Institute on Mental
Health/RAND), and "An Ethnographic Analysis of Multi-Problem Street Youth
at Risk for AIDS" (funded by the National Institute on Drug Abuse).

Ann Lessem, Ph.D. is an assistant research scientist at the Public Policy Research Institute at Texas A&M University. She received her doctorate in educational administration from Texas A&M. Dr. Lessem serves as a principal investigator for PPRI research and evaluation projects. She is involved with proposal
writing, instrument design, data collection and analysis, and report preparation.
She has conducted research in the fields of education, workforce development,
labor market analysis, and economic development in rural communities. With
expertise in ethnographic and qualitative/field generated data collection and
analysis, Dr. Lessem has also been involved in a National Institute for Drug
Abuse cyber-ethnography of club drug culture.

Edward Murguía, Ph.D. is an associate professor of sociology at Texas A&M
University. He received his doctorate from the University of Texas at Austin in
1978. He was a Minority Scholar, National Institute of Drug Abuse from 1993
to 1997, as well as a Research Associate, Laboratory for the Studies of Social
Deviance, Texas A&M University, during those same years. His article entitled,
"A Comparison of Causal Factors in Drug Use among Mexican Americans and
Non-Hispanic Whites," appeared in *Social Science Quarterly* in 1998. He was
primary investigator of a NIDA grant in 2001 to study youth, the Internet, and
drug use, and presented some of his findings at the National Institute of Drug
Abuse in Rockville, MD, in 2002 and 2003.

Melissa Tackett-Gibson, Ph.D. is an assistant research scientist at the Public
Policy Research Institute at Texas A&M University. She received her doctorate
in sociology from Northeastern University in 2002. Dr. Tackett-Gibson serves as
a principal investigator for PPRI research and evaluation projects, and since
1997, she has been involved in the development and implementation of health
and policy related research. Dr. Tackett-Gibson has a background in substance

abuse research methodologies and the social impact of abuse. In addition to "Youth, Technology, and the Proliferation of Drug Use," she is involved in the analyses of substance abuse data for the Texas Council on Alcohol and Drug Abuse funded, "Texas School Survey of Substance Abuse." She is currently focusing on the misuse and abuse of prescription medications and correlated risk behaviors.

Index